Finding the Forest in the Trees

The Challenge of Combining Diverse Environmental Data

Selected Case Studies

Committee for a Pilot Study on
Database Interfaces

U.S. National Committee for CODATA

Commission on Physical Sciences, Mathematics, and Applications
National Research Council

NATIONAL ACADEMY PRESS
Washington, D.C. 1995

NOTICE: The project that is the subject of this report was approved by the Governing Board of the National Research Council, whose members are drawn from the councils of the National Academy of Sciences, the National Academy of Engineering, and the Institute of Medicine. The members of the committee responsible for the report were chosen for their special competences and with regard for appropriate balance.

This report has been reviewed by a group other than the authors according to procedures approved by a Report Review Committee consisting of members of the National Academy of Sciences, the National Academy of Engineering, and the Institute of Medicine.

Support for this project was provided by the Department of Energy (under Grant No. DE-FG-02-92ER61514); the Environmental Protection Agency (under Contract No. 68D10096); the National Aeronautics and Space Administration (under Contract No. W88958); the National Institute of Standards and Technology (under Contract No. 50SBNB3C7500); and the National Oceanic and Atmospheric Administration (through Grant No. INT-9118880). Although the results described in this document have been funded in part by the United States Environmental Protection Agency under the Assistance Agreement to the National Academy of Sciences, the document has not been subjected to the Agency's peer and administrative review and therefore may not necessarily reflect the views of the Agency, and no official endorsement should be inferred. Any opinions, findings, conclusions, or recommendations expressed in this material are those of the authors and do not necessarily reflect the views of the sponsoring agencies.

Library of Congress Catalog Card Number 95-67403
International Standard Book Number 0-309-05082-0

Additional copies of this report are available from:

National Academy Press
2101 Constitution Avenue, NW
Box 285
Washington, DC 20055
800-624-6242
202-334-3313 (in the Washington Metropolitan Area)

B-345

Copyright 1995 by the National Academy of Sciences. All rights reserved.

Printed in the United States of America

COMMITTEE FOR A PILOT STUDY ON DATABASE INTERFACES

G. BRUCE WIERSMA, University of Maine, *Chair*
SHELTON A. ALEXANDER, Pennsylvania State University
DELBERT BARTH, University of Nevada, Las Vegas
MARION BAUMGARDNER, Purdue University
BROCK BERNSTEIN, EcoAnalysis, Inc.
DALE A. BRUNS, Wilkes University
ALI H. GHOVANLOU, MITRE Corporation
ROBERT E. MUNN, University of Toronto
ESTELLE RUSSEK-COHEN, University of Maryland

Staff

Paul F. Uhlir, Associate Executive Director, Commission on Physical Sciences, Mathematics, and Applications
Julie M. Esanu, Research Assistant
David J. Baskin, Project Assistant

U.S. NATIONAL COMMITTEE FOR CODATA

DAVID R. LIDE, Jr., National Institute of Standards and Technology (retired), *Chair*
STANLEY M. BESEN, Charles River Associates, Inc.
LOIS D. BLAINE, American Type Culture Collection
ALI H. GHOVANLOU, MITRE Corporation
JULIAN HUMPHRIES, Cornell University
MICAH I. KRICHEVSKY, Bionomics International
KENNETH N. MARSH, Texas A&M University
GOETZ OERTEL, Association of Universities for Research in Astronomy
STANLEY RUTTENBERG, University Corporation for Atmospheric Research (retired)
PAMELA SAMUELSON, University of Pittsburgh
JACK H. WESTBROOK, Brookline Technologies

Staff

Paul F. Uhlir, Associate Executive Director, Commission on Physical Sciences, Mathematics, and Applications
Julie M. Esanu, Research Assistant
David J. Baskin, Project Assistant

COMMISSION ON PHYSICAL SCIENCES, MATHEMATICS, AND APPLICATIONS

RICHARD N. ZARE, Stanford University, *Chair*
RICHARD S. NICHOLSON, American Association for the Advancement of Science, *Vice Chair*
STEPHEN L. ADLER, The Institute for Advanced Study
SYLVIA T. CEYER, Massachusetts Institute of Technology
SUSAN L. GRAHAM, University of California at Berkeley
ROBERT J. HERMANN, United Technologies Corporation
RHONDA J. HUGHES, Bryn Mawr College
SHIRLEY A. JACKSON, Rutgers University
KENNETH I. KELLERMANN, National Radio Astronomy Observatory
HANS MARK, University of Texas at Austin
THOMAS A. PRINCE, California Institute of Technology
JEROME SACKS, National Institute of Statistical Sciences
L.E. SCRIVEN, University of Minnesota
A. RICHARD SEEBASS III, University of Colorado
LEON T. SILVER, California Institute of Technology
CHARLES P. SLICHTER, University of Illinois at Urbana-Champaign
ALVIN W. TRIVELPIECE, Oak Ridge National Laboratory
SHMUEL WINOGRAD, IBM T.J. Watson Research Center
CHARLES A. ZRAKET, MITRE Corporation (retired)

NORMAN METZGER, Executive Director
PAUL F. UHLIR, Associate Executive Director

The National Academy of Sciences is a private, nonprofit, self-perpetuating society of distinguished scholars engaged in scientific and engineering research, dedicated to the furtherance of science and technology and to their use for the general welfare. Upon the authority of the charter granted to it by Congress in 1863, the Academy has a mandate that requires it to advise the federal government on scientific and technical matters. Dr. Bruce Alberts is president of the National Academy of Sciences.

The National Academy of Engineering was established in 1964, under the charter of the National Academy of Sciences, as a parallel organization of outstanding engineers. It is autonomous in its administration and in the selection of its members, sharing with the National Academy of Sciences the responsibility for advising the federal government. The National Academy of Engineering also sponsors engineering programs aimed at meeting national needs, encourages education and research, and recognizes the superior achievements of engineers. Dr. Robert M. White is president of the National Academy of Engineering.

The Institute of Medicine was established in 1970 by the National Academy of Sciences to secure the services of eminent members of appropriate professions in the examination of policy matters pertaining to the health of the public. The Institute acts under the responsibility given to the National Academy of Sciences by its congressional charter to be an adviser to the federal government and, upon its own initiative, to identify issues of medical care, research, and education. Dr. Kenneth I. Shine is president of the Institute of Medicine.

The National Research Council was organized by the National Academy of Sciences in 1916 to associate the broad community of science and technology with the Academy's purposes of furthering knowledge and advising the federal government. Functioning in accordance with general policies determined by the Academy, the Council has become the principal operating agency of both the National Academy of Sciences and the National Academy of Engineering in providing services to the government, the public, and the scientific and engineering communities. The Council is administered jointly by both Academies and the Institute of Medicine. Dr. Bruce Alberts and Dr. Robert M. White are chairman and vice chairman, respectively, of the National Research Council.

Preface

The U.S. National Committee (USNC) for the Committee on Data for Science and Technology (CODATA) created the Committee for a Pilot Study on Database Interfaces (the committee) to review and advise on data interfacing activities because of the increasing importance of conducting interdisciplinary environmental research and assessments, both nationally and internationally. CODATA is an interdisciplinary body of the International Council of Scientific Unions concerned with all types of quantitative data resulting from experimental measurements or observations in the physical, biological, geological, and astronomical sciences. The USNC/CODATA is a standing committee organized under the National Research Council (NRC) to represent U.S. interests in the international CODATA.

The charge to the committee from the USNC/CODATA was as follows:

> The committee will identify data management problems and issues associated with physical, biological, and chemical parameters that are important to interdisciplinary environmental research, particularly as it relates to long-term global change studies such as the International Geosphere-Biosphere Program. The committee also will look at reducing barriers that may occur in interfacing biological and ecological databases with geophysical and remote sensing databases. The committee will propose guidelines for interfacing interdisciplinary environmental data.

The report does not address in detail the mathematical and statistical aspects associated with data interfacing activities, which were the topic of a recent NRC report, *Combining Information: Statistical Issues and Opportunities for Research* (National Academy Press, Washington, D.C., 1992). Nor does it address the issue of technical barriers in the electronic storage and distribution of interdisciplinary environmental data. Rather, the focus is on helping researchers and technicians engaged in interdisciplinary research, particularly those projects that involve both geophysical and ecological issues, to better plan and implement their supporting data management activities. It also is aimed at informing those individuals responsible for funding, managing, or evaluating such studies and activities.

G. Bruce Wiersma, *Chair*
Committee for a Pilot Study on
Database Interfaces

Acknowledgments

The committee for a Pilot Study on Database Interfaces used six case studies to derive the conclusions and recommendations presented in this report.

For the information they provided on the Impact Assessment Project for Drought Early Warning in the Sahel, the case study described in Chapter 2, the committee is grateful to Gerald Barton, Susan Callis, Douglas LeComte, and Sharon LeDuc of NOAA.

The committee began its review of the Aquatic Processes and Effects portion of the National Acid Precipitation Assessment Program (NAPAP), the case study described in Chapter 3, by interviewing or obtaining presentations from key Environmental Protection Agency (EPA) participants. EPA scientists also suggested additional individuals who could provide more specific information on selected aspects of the studies. Patricia Irving, executive director of NAPAP at the Council on Environmental Quality (CEQ) until April 1992 and Jack Barnes of her staff were contacted to obtain all the relevant references.

On a February 1992 site visit to the EPA Research Laboratory at Corvallis, Oregon, the committee interviewed Dixon Landers, M. Robbins Church, Jeffrey Lee, and Parker J. Wigington—all of whom played key roles in EPA's implementation of the NAPAP *Research Plan on Aquatic Processes and Effects* (Interagency Task Force on Acid Precipitation, Washington, D.C., 1982). The committee subsequently obtained presentations from Robb Turner, Oak Ridge National Laboratory, on the subject of data management, and from Joan Baker, the Cadmus Group, regarding the biological effects of acid precipitation. Derek Winstanley, the executive director of NAPAP after Dr. Irving, also briefed the committee on plans to continue NAPAP, as authorized by the Clean Air Act Amendments of 1990.

Finally, the chair of the subcommittee for this case study conducted informal interviews at EPA's Environmental Monitoring Systems Laboratory in Las Vegas, Nevada, with key scientists who were involved with the implementation of EPA's portion of NAPAP. These included Gene Meier, Robert Schonbrod, Daniel Heggem, and Louis Blume.

The Long-Term Ecological Research (LTER) subcommittee met in Corvallis, Oregon, on November 5-6, 1992, to gather information for the case study on the H.J. Andrews Experimental Forest LTER site, described in Chapter 4. Members were briefed on the first day by the Andrews Forest staff and by the faculty of the College of Forestry at Oregon State University (OSU). The individuals providing the briefings and input were Susan Stafford, director of the Quantitative Sciences Group; Hazel Hammond, GIS software specialist; Donald Henshaw, U.S. Forest Service database manager; Gody Spycher, OSU database manager; Rudolf Nottrott, LTER database manager; and Arthur McKee, director, Andrews Forest and LTER principal co-investigator. On the second day the committee went with Arthur McKee, Donald Henshaw, and Rudolf Nottrott for a site visit to the Andrews Forest.

Robert Cushman presented a briefing on the Carbon Dioxide Information Analysis Center (CDIAC), the subject of Chapter 5, to the committee in Washington, D.C., on March 30, 1992. Following this, several members of the committee met in Oak Ridge, Tennessee, on May 18-19, 1992, to gather more detailed information about the Center's data management and analysis activities. Briefings were carried out by ORNL and CDIAC staff—Paul Kanciruk, Thomas Boden, Robert Cushman, and Frederick Stoss. These covered the history of the Center and its management structure, the hardware and software systems currently being used, the types of data packages produced, and the philosophy and procedures underlying the Center's work in managing and integrating the data.

In November 1992, the committee was briefed by Forrest G. Hall of the NASA Goddard Space Flight Center in Greenbelt, Maryland, and Donald E. Strebel of Versar, Inc. in Columbia, Maryland, and received extensive documentation on the First ISLSCP Field Experiment (FIFE), the case study described in Chapter 6. In particular, the FIFE scientists prepared very helpful written responses to a series of specific questions posed in advance by the committee. The committee also acknowledges with thanks the information on FIFE that it received through R.E. Munn from W. Brutsaert (Cornell University), R.L. Desjardins (Agriculture Canada), and J. Kay (University of Waterloo).

ACKNOWLEDGMENTS

In Chapter 7, the California Cooperative Oceanic Fisheries Investigation (CalCOFI) case study draws on written material prepared by the Panel on the Southern California Bight of the NRC's Committee on a Systems Assessment of Marine Environmental Monitoring. It also benefited from conversations with John McGowan of Scripps Institution of Oceanography and George Hemingway, CalCOFI coordinator.

Contents

EXECUTIVE SUMMARY		1
1	INTRODUCTION	12
	Methodology of the Study, 14	
	Organization of the Report, 17	
	References, 17	
2	IMPACT ASSESSMENT PROJECT FOR DROUGHT EARLY WARNING IN THE SAHEL	18
	Variables Measured and Sources of Data, 20	
	Data Management and Interfacing, 22	
	Lessons Learned, 27	
	References, 29	
3	THE NATIONAL ACID PRECIPITATION ASSESSMENT PROGRAM	30
	Variables Measured and Sources of Data for the Aquatic Processes and Effects Portion of NAPAP, 31	
	Major Considerations in Evaluating the Data Management Activities of the Aquatic Processes and Effects Portion of NAPAP, 34	
	Data Management and Interfacing in the Aquatic Processes and Effects Portion of NAPAP, 38	
	Lessons Learned, 45	
	References, 45	

xiii

| 4 | THE H.J. ANDREWS EXPERIMENTAL FOREST LONG-TERM ECOLOGICAL RESEARCH SITE | 46 |

Variables Measured and Sources of Data, 48
Data Management and Interfacing, 51
Lessons Learned, 54
References, 55

| 5 | THE CARBON DIOXIDE INFORMATION ANALYSIS CENTER | 56 |

Variables Measured and Sources of Data, 57
Data Management, 58
Lessons Learned, 62
References, 65

| 6 | THE FIRST ISLSCP FIELD EXPERIMENT | 66 |

Variables Measured and Sources of Data, 67
Data Management and Interfacing, 68
Lessons Learned, 71
References, 72

| 7 | THE CALIFORNIA COOPERATIVE OCEANIC FISHERIES INVESTIGATION | 74 |

Variables Measured and Sources of Data, 75
Data Management and Interfacing, 76
Lessons Learned, 79
References, 80

| 8 | INTERFACING DIVERSE ENVIRONMENTAL DATA— ISSUES AND RECOMMENDATIONS | 81 |

The Problem and Its Context, 82
Addressing Barriers Deriving from the Data, 84
Addressing Barriers Deriving from Users' Needs, 93
Addressing Barriers Deriving from Organizational Interactions, 97
Addressing Barriers Deriving from Information System Considerations, 103
Ten Keys to Success, 112
References, 116

APPENDIXES

| A | Case Study Evaluation Criteria | 121 |
| B | List of Abbreviations and Acronyms | 128 |

Executive Summary

We are the first generation to have the tools to study the Earth as a system. During the last few decades of the 20th century, the development of an array of technologies has made it possible to observe the Earth, collect large quantities of data related to components and processes of the Earth system, and store, analyze, and retrieve these data at will. These data can be registered to specific locations on the Earth's surface and can be integrated into spatial-temporal information systems and registered at the same scale and cartographic projection as other resource data.

Scientists can now perform environmental research that increases our understanding of the Earth system at all spatial scales, enhances resource management and environmental decision making, and improves our capabilities for predicting significant changes in the environment. Over the past decade, in particular, the observational, computational, and communications technologies have enabled the scientific community to undertake a broad range of interdisciplinary environmental research and assessment programs. At the international level, two of the most ambitious programs are the International Geosphere-Biosphere Program (IGBP) of the International Council of Scientific Unions (ICSU), and the World Climate Research Program, jointly sponsored by the World Meteorological Organization and ICSU. At the national level, these international research initiatives are supported through the federal interagency Global Change Research Program.

Global change research, by its nature and scope, is inherently complex. On the technical side, complexity increases with the number of different variables that are modeled, measured, or experimentally manipulated. These variables may interact with each other to a high degree, and these interactions include nonlinearities or discontinuities in space or time. In particular, a certain degree of complexity in global change research ensues from the sheer quantity of data at large spatial and temporal scales. Likewise, analogous degrees of complexity originate on the organizational side of research in how the work is structured, managed, and implemented due to the sizable number of investigators and participants across a range of disciplines.

The Global Change Research Program and other large research initiatives involve the interfacing of large volumes of diverse data, commonly combining several traditionally distinct disciplines, such as meteorology, oceanography, geology, biology, chemistry, and geography, or their related subdisciplines. "Data interfacing" may be defined as the coordination, combination, or integration of data for the purpose of modeling, correlation, pattern analysis, hypothesis testing, and field investigations at various scales. Because data from each discipline and subdiscipline are organized into data sets and databases that frequently possess unique or special attributes, their effective interfacing can be difficult.

Sound practices in database management are required to deal effectively with problems of complexity in global change studies and other large interdisciplinary research and assessment projects. Although a great deal of attention and resources has been devoted to this type of research in recent years, little guidance has been provided on overcoming the barriers frequently encountered in the interfacing of disparate data sets. And although there is a wealth of relevant experience at the working level in the research community, this experience generally has not been analyzed and organized to make it more readily available to researchers.

Because of the increasing importance of conducting interdisciplinary environmental research and assessments, both nationally and internationally, the Committee for a Pilot Study on Database Interfaces was charged to review and advise on data interfacing activities in that context. This report is the result of that study. The focus is on developing analytical and functional guidelines to help researchers and technicians engaged in interdisciplinary research—particularly those projects that involve both geophysical and ecological issues—to better plan and implement their supporting data management activities. It also is aimed at informing those individuals responsible for funding, managing, or evaluating such studies and activities.

SUMMARY OF CONCLUSIONS AND RECOMMENDATIONS

The committee used six case studies (1) to identify and to understand the most important problems associated with collecting, integrating, and analyzing environmental data from local to global spatial scales and over a very wide range of temporal scales, and (2) to elaborate the common barriers to interfacing data of disparate sources and types. Consistent with the committee's charge, the primary focus was on the interfacing of geophysical and ecological data. The committee derived a number of lessons from the case studies, and these lessons are summarized at the end of each case study and analyzed in Chapter 8. Some are generic in nature; others are more specific to a discipline or project.

The conclusions and recommendations are all based on the committee's analysis of the case studies and on additional research. They are organized according to four major areas of barriers or challenges to the effective interfacing of diverse environmental data. These are barriers deriving from the data themselves, from the users' needs, from organizational interactions, and from system considerations. In the final section the committee offers a set of broadly applicable principles—Ten Keys to Success—that can be used by scientists and data managers in planning and conducting data interfacing activities.

Addressing Barriers Deriving from the Data

The spatial and temporal scales of the disciplines important to environmental research vary enormously. Such variation was certainly typical of the ecological and geophysical data sets that were reviewed in this study. For instance, massive data sets that cover large areas are routinely collected and used in the physical sciences, while such data sets are much less common in ecological disciplines. These differences reflect distinct historical traditions, working methods, and judgments about what processes are important and the temporal and spatial scales on which they operate. As a result, it is difficult to find geophysical and ecological data sets with matching temporal and spatial scales. In addition, attempts to equalize scales through various methods of data manipulation run the risk of creating spurious patterns and correlations.

Recommendation 1. In the planning for interdisciplinary research, careful thought should be given to the implications of different inherent spatial and temporal scales and the processes they represent. These should be discussed explicitly in project planning documents. The methods used to accommodate or match inherent scales in different data types in any attempts to facilitate modeling and analysis should be carefully evaluated for their potential to produce artificial patterns and correlations.

Preliminary processing generally is necessary to develop useful derived data products from raw data. As a result, data sets unavoidably reflect certain scientific assumptions, perspectives, and value judgments. In addition, each processing step is associated with some kind and amount of statistical uncertainty.

Recommendation 2. Metadata* should explicitly describe all preliminary processing associated with each data set, along with its underlying scientific purpose and its effects on the suitability of the data for various purposes. Further, metadata also should describe and quantify to the extent feasible the statistical uncertainty resulting from each processing step. Planning for studies that involve interfacing should explicitly consider the effects of preliminary processing on the utility of the resultant integrated data set(s). (Additional recommendations regarding metadata appear below.)

The exceptionally large data volumes involved in global environmental research can pose significant challenges for existing methods of data storage, retrieval, and analysis, as well as for the organizational systems currently in place to support these activities.

Recommendation 3. All proposed data management and interfacing methods should be weighed carefully in terms of their ability to deal with large volumes of data. Assumptions that existing methods will continue to be suitable should be treated with caution.

The committee found that differences in scientific conventions among disciplines can be a severe impediment to data interfacing, significantly increasing the costs of achieving compatibility among data sets and in some cases preventing it completely. Some of these differences stem from fundamental dissimilarities in study design or purpose and others from traditional practice that varies from discipline to discipline.

Recommendation 4. Efforts to establish data standards should focus on a key subset of common parameters whose standardization would most facilitate data interfacing. Where possible, such standardization should be addressed in the initial planning and design phases of interdisciplinary research. Early attention to integrative modeling can help identify key incompatibilities. The data requirements, data characteristics and quality, and scales of measurement and sampling should be well defined at the outset.

In several of the case studies, essential ecological data sets were either missing or of unknown quality. In some cases, it was necessary to create

---

*The committee defines metadata as the documentation or description of the facts, circumstances, and conditions associated with the actual data themselves. This term is used interchangeably with "documentation" throughout this text.

such needed data sets by using historical data or by combining data from a range of ecological studies.

Recommendation 5. Agencies that perform or support environmental research and assessment generally, and global change research particularly, should identify and define key ecological data sets that do not exist but are important to their mission. A careful review should be made of options for finding, rescuing, or creating these crucial data, and funding should be set aside to implement the most feasible option(s).

Addressing Barriers Deriving from Users' Needs

Users' needs in global change research are exceptionally diverse, fluid, and difficult to predict. These characteristics require that data management systems and practices be designed for maximum ease of access, adaptability over time, and communication among all potential users. However, the committee concludes that existing practices frequently inhibit communication and exchange of ideas with the larger user community, as well as access to the data by secondary and tertiary users.

Recommendation 6. Project scientists and data managers should adopt the view that one of their primary responsibilities is the creation of long-lasting data and information resources for the broad research community. Data management systems and practices, particularly the development of metadata, should be designed to balance the needs of this larger user community with those of project scientists.

Addressing Barriers Deriving from Organizational Interactions

The committee concludes that the existing missions and attendant reward systems of research organizations act to inhibit the data sharing, mutual support, and interdisciplinary mindset needed for successful data interfacing. In many cases the stated aims of global change research programs are at odds with the collective understanding among staff within organizations of what their job responsibilities are and how they should be fulfilled.

Recommendation 7. Professional societies, research institutions, and funding and management agencies should reevaluate their reward systems in order to give deserved peer recognition to scientists and data managers for their contributions to interdisciplinary research. Granting and funding agencies, as well as program managers and university administrators, should provide tangible incentives to motivate scientists to participate actively in data management and data interfacing activities. Such incentives should extend to favorable consideration of those activities in performance reviews, including treating the production of value-added data sets as analogous to scientific publications.

Recommendation 8. Because organizational missions and reward systems inherently reflect a larger policy context, relevant policy issues should be included in the planning for interdisciplinary research. This should be accomplished in part through open communication between project scientists and appropriate policymakers that continues throughout the life of the project. Such communication will help provide a basis for developing cooperative arrangements between collaborating institutions that will provide strong incentives for and reduce barriers to sharing data.

The case studies considered by the committee covered a broad range of objectives, spatial and temporal scales, data sources, data management procedures, quantity and quality of data, and analytic and interpretive methods. From these observations and a consideration of the results of the case studies, the committee concludes that effective data management is an integral part of successful data interfacing. The committee also concludes that there is a critical need to educate scientists about data management principles and to foster improved working relationships between scientists and information management professionals.

Recommendation 9. Research universities should include courses in their curricula that provide environmental scientists with an in-depth understanding of the rationale for and principles of sound data management. Program managers and data managers, in their interactions with and training of environmental scientists, should emphasize how state-of-the-practice data management can provide immediate and long-lasting benefits to scientists, particularly those engaged in interdisciplinary research. At the same time, data managers need to be a part of the conceptual team from the beginning of a project and have equal status with principal investigators.

In its review of the factors that contribute to the success or failure of data interfacing efforts, the committee identified traditional concepts of data ownership as a serious impediment to success. Existing reward systems and traditional practice often combine to motivate scientists to treat data as personal property, even in the face of contractual agreements for data submission and sharing.

Recommendation 10. In order to encourage interdisciplinary research and to make data available as quickly as possible to all researchers, specific guidelines should be established for when and under what conditions data will be made available to users other than those who collected them. Such guidelines are particularly important when data collectors, data managers, and other users are in different organizations. In addition, adequate rewards should be established by the funders of research and publishers to motivate principal investigators to place all data in the public domain.

A major factor in planning for successful data interfacing is the choice of personnel and the institutional arrangements in which they work. The committee found many instances in which optimal interdisciplinary activities and data sharing were not possible because of unclear responsibilities, conflicting goals, misunderstanding, and outright rivalries. The added complexity of interdisciplinary research increases the severity of such common organizational problems. Even one organization or key player who refuses to share data, prepare documentation, participate in standards setting, or provide other vital project support in a collaborative effort can greatly diminish the probability of success.

Recommendation 11. In the planning of any interdisciplinary research program, as much consideration should be given to organizational and institutional issues as to technical issues. Every effort should be made to minimize the likelihood of misunderstanding, conflicts, and rivalries by establishing interorganizational relationships and procedures, creating effective reward structures, and creating new functions that explicitly support data interfacing.

Based on the case studies and related research, the committee concludes that insufficient attention is given in many interdisciplinary studies to quality control, beta testing of derived data products, creation of broadly useful value-added data sets, resolution of data compatibility problems, and the maintenance and security of key data sets on a long-term basis. Many of these functions are beyond the scope envisioned for existing data centers.

Recommendation 12. The agencies involved in supporting and carrying out interdisciplinary research should investigate the possibility of establishing one or more ecosystem data and information analysis centers to facilitate the exchange of data and access to data, help improve and maintain the quality of valuable data sets, and provide value-added services. A model for such a center is the Carbon Dioxide Information Analysis Center (CDIAC) at Oak Ridge National Laboratory. In addition, it would be wise to look closely at the potential synergism between any new ecosystem data and information analysis center and all other existing environmental data centers.

Addressing Barriers Deriving from System Considerations

The nature of interdisciplinary global change research makes it impossible to clearly define a detailed and stable set of user requirements. Classical methods of system design are therefore inappropriate because they do not provide for sufficient user input throughout the entire design effort and do not incorporate adequate provisions for flexibility and adaptability.

Recommendation 13. **Hardware/software system development efforts should be based on a model that includes ongoing interaction with users as an integral part of the design process. In addition, system designers should work from the assumption that systems will never be finished, but will continue to evolve along with the data collected and users' needs. Designers therefore should use, to the greatest extent possible, modern database development approaches such as rapid prototyping, modular systems design, and object-oriented programming, which enhance system adaptability.**

One of the conclusions that clearly emerged from the case studies is the critical role of system interoperability in supporting data interfacing efforts. Interoperability is the ability to readily connect different databases on separate hardware and software systems and perform data retrieval, analyses, and other applications without regard to the boundaries between the systems. Given current technology, this can be a difficult goal to achieve, and it currently requires the direct involvement of knowledgeable information management specialists. However, even when hardware and software systems are successfully connected, fundamental incompatibilities among the data themselves can still impede interfacing.

Recommendation 14. **Program managers, project scientists, and data managers should review the interoperability of their hardware, software, and data management technologies to facilitate locating, retrieving, and working with data across several disciplines. However, this effort should be accompanied by parallel attempts to resolve inherent incompatibilities among data types that can thwart interfacing even when state-of-the-art hardware and software systems are seamlessly connected.**

One of the most serious problems in the creation, integration, use, and management of large databases for interdisciplinary research is the lack of adequate metadata. Metadata enable users other than the principal investigator to make effective use of the data and to determine which applications they may or may not be suited for. The committee found instances in the case studies where data sets had to be discarded because investigators did not provide the documentation needed for others to make use of them. It is important for researchers to understand that the incremental cost of including the necessary documentation at the time of data collection is small in comparison with the cost of attempting to reconstruct it retrospectively at the end of the project, or long after it has been completed, which may be prohibitive in cost or impossible to do.

Recommendation 15. **The production of detailed metadata should be a mandatory requirement of every study whose data might be used**

for interdisciplinary research. Metadata should be treated with the seriousness of a peer-reviewed publication and should include, at a minimum, a description of the data themselves, the study design and data collection protocols, any quality control procedures, any preliminary processing, derivation, extrapolation, or estimation procedures, the use of professional judgment, quirks or peculiarities in the data, and an assessment of features of the data that would constrain their use for certain purposes.

The committee found that interdisciplinary research almost invariably involves using data in ways not initially envisioned by the original investigators. In many cases, new uses of data require backtracking along the data path in order to reformat, resummarize, reclassify, or otherwise adjust the data to make them suitable for current needs. In order to backtrack, detailed information should be available about the prior processing steps that were used to create the data sets being interfaced.

Recommendation 16. Metadata should contain enough information to enable users who are not intimately familiar with the data to backtrack to earlier versions of the data so that they can perform their own processing or derivation as needed. Where stand-alone documentation is not adequate (for large and complex data sets or where multiple users are simultaneously updating and modifying data), data managers should investigate the feasibility of incorporating an audit trail into the data themselves.

The committee concludes that far too many environmental research projects give insufficient attention, in either the planning or the implementation stage, to the long-term archiving of their data sets. Data from studies that contribute significantly to our understanding of components and processes of the Earth system must be preserved and made accessible for future potential users of the data. There is a need to create a mindset within all elements of the research community that valuable data need to have a long-term life that extends far beyond the publication of the principal investigator's analyses.

Recommendation 17. In general, the presumption in environmental research should be that "data worth collecting are worth saving." Funding agencies therefore should consider stipulating that all research applicants include in their research plans well-conceived and adequately funded arrangements for data management and for the ultimate disposition of their data. While it is impossible to establish universal guidelines for funding, the committee's investigations suggest that setting aside 10 percent of the total project cost for data management would not be unreasonable. These cost estimates should include adequate funds for preparing thorough metadata that serve the needs of all potential users. In order for these requirements to be fully

effective, however, the agencies must adequately support active archives and long-term data repositories. (See also Recommendation 12.)

There are no well-established and widely accepted protocols to assist scientists in deciding which data should be archived, in what formats they should be stored, and where and how they should be archived to maximize access for potential future users. Further, in several cases the committee found little attention given to the long-term maintenance of data sets once they were archived. It is important to note, however, that there are no technical barriers to keeping all data collected in research projects, even data-intensive ones that involve high-resolution imagery, because advances in data storage and retrieval capabilities have kept pace with the ever-growing volumes of data in all fields of science. It is typical that the ensemble of all previous data in any scientific discipline is modest in volume compared to present and anticipated annual volumes. Therefore, the issue is not unmanageable volumes of data; rather it is the maintenance of the data sets in accessible, usable form over time that is the challenge for long-term retention.

Recommendation 18. The committee is concerned about the gaps in the existing system for long-term retention and maintenance of environmental data. Funding agencies should provide guidelines that define the requirements for preparing data sets for long-term archiving. Educational and research institutions should be encouraged to incorporate strong data management and archival activities into every interdisciplinary project and should allocate sufficient funding to accomplish these functions. Professional recognition should be given to principal investigators and project data managers who perform these functions well.

TEN KEYS TO SUCCESS

The committee's investigations of the case studies and other related experience identified Ten Keys to Success, each of which incorporates both technical and cultural aspects. Keys 1 and 2 deal with the appropriate use of available information management technology. Keys 3, 4, and 5 describe design and management strategies. Keys 6, 7, and 8 refer to methods for accommodating the unavoidable realities of human behavior, motivation, and politics. Finally, keys 9 and 10 suggest ways of enhancing data interfacing by building the need for it into the structure of research programs.

1. Be practical.
2. Use appropriate information technology.
3. Start at the right scale.

EXECUTIVE SUMMARY

4. Proceed incrementally.
5. Plan for and build on success.
6. Use a collaborative approach.
7. Account for human behavior and motivation.
8. Consider needs of participants as well as users.
9. Create common needs for data.
10. Build participation by demonstrating the value of data interfacing.

1

Introduction

We are the first generation to have the tools to study the Earth as a system. During the last few decades of the 20th century, the development of an array of technologies has made it possible to observe the Earth, collect large quantities of data related to components and processes of the Earth system, and store, analyze, and retrieve these data at will. These data can be registered to specific locations on the Earth's surface and can be integrated into spatial-temporal information systems and registered at the same scale and cartographic projection as other resource data.

Another important technological advance, the Internet, has had a major impact on the nature of all scientific research. This electronic network links computers located in universities, government agencies and laboratories, as well as many commercial enterprises. The network, initially developed and partially supported by the federal government, allows rapid communication among scientists. Many groups have set up Wide Area Information Servers (WAIS), which allow users at other nodes in the network to retrieve data. These developments have encouraged researchers to retrieve and combine data sets in ways not previously attempted.

Scientists can now perform environmental research that increases our understanding of the Earth system at all spatial scales, enhances resource management and environmental decision making, and improves our capabilities for predicting significant changes in the environment. Over the past decade, in particular, these observational, computational, and communications technologies have enabled the scientific community to undertake a broad range of interdisciplinary environmental research and assessment programs. At the international level, two of the most ambitious programs are the International Geosphere-Biosphere Program (IGBP) of the International Council of Scientific Unions (ICSU), and the World Climate Research Program, jointly sponsored by the World Meteorological Organization and ICSU (NRC, 1990). At the national level, these international research initiatives are supported through the federal interagency Global Change Research Program (NSTC, 1994).

Global change research, by its nature and scope, is inherently complex. On the technical side, complexity increases with the number of different variables that are modeled, measured, or experimentally manipulated. These variables may interact with each other to a high degree, and these interactions include nonlinearities or discontinuities in space or time. In particular, a certain degree of complexity in global change research ensues from the sheer quantity of data at large spatial and temporal scales. Likewise, analogous degrees of complexity might originate on the organizational side of research in how the work is structured, managed, and implemented due to the sizable number of investigators and participants across a range of disciplines.

The Global Change Research Program and other large research initiatives involve the interfacing of large volumes of diverse data, commonly combining several traditionally distinct disciplines, such as meteorology, oceanography, geology, biology, chemistry, and geography, or their related subdisciplines. "Data interfacing" may be defined as the coordination, combination, or integration of data for the purpose of modeling, correlation, pattern analysis, hypothesis testing, and field investigations at various scales. Because data from each discipline and subdiscipline are organized into data sets and databases that frequently possess unique or special attributes, their effective interfacing can be difficult.

Sound practices in database management are required to deal effectively with problems of complexity in global change studies and other large interdisciplinary research and assessment projects. Although a great deal of attention and resources has been devoted to this type of research in recent years, little guidance has been provided on overcoming the barriers frequently encountered in the interfacing of disparate data sets. And although there is a wealth of relevant experience at the working level in the research community, this experience generally has not been analyzed and organized to make it more readily available to researchers.

Because of the increasing importance of conducting interdisciplinary environmental research and assessments, both nationally and internationally, the Committee for a Pilot Study on Database Interfaces was charged to review and advise on data interfacing activities in that context. This report is the result of that study. It does not address in detail the mathematical and statistical aspects associated with data interfacing activities, which were the topic of a recent NRC report (see NRC, 1992). Nor does it address the issue of technical barriers in the electronic storage and distribution of interdisciplinary environmental data. Rather, the focus is on developing analytical and functional guidelines to help researchers and technicians engaged in interdisciplinary research—particularly those projects that involve both geophysical and ecological issues—to better plan and implement their supporting data management activities. It also is aimed at informing those individuals responsible for funding, managing, or evaluating such studies and activities.

METHODOLOGY OF THE STUDY

Early in its deliberations the committee decided to take a case study approach. The objective was to obtain some well-documented examples of successful and unsuccessful data interfacing techniques and to learn from the triumphs and failures of those who had actually conducted complex interdisciplinary environmental studies. One drawback to this approach was that there still are not many completed studies that have focused on global environmental change. Another limitation was that none of the case studies used the Internet as a key technological component. Nevertheless, the committee believes that the rapidly increasing use of the Internet in many research projects will accentuate the data management issues examined in this report.

Because the committee wanted to select the maximum number of diverse case studies feasible to examine in the time available, it identified 11 potential cases for consideration. It was particularly interested in complex, interdisciplinary studies in which a combination of physical, chemical, and biological measurements were taken and then integrated into composite data sets from which various conclusions could be drawn. The focus was on evaluating the data management activities in each case study rather than the research itself.

The committee used a modified delphi technique—a method of quickly quantifying and weighing diverse opinions—to rank the candidate case studies. The following criteria were used in deciding which case studies to select:

• If possible, the studies should involve global change research or assessment.
• The research should be interdisciplinary.
• There should be reasonable access to the results of the studies and to their designers.

- The studies should involve some attempt at integrating multimedia data sets from both the geophysical and the biological sciences.
- The studies should be spread over a variety of different scales of activities and operations, from international to local, and from large-scale to small-scale.
- The studies should cover a wide range of environmental issues.
- The studies should either involve completed research projects or projects that have been under way long enough to have developed and used complex data management systems.

On the basis of these criteria the committee selected the following six case studies for detailed investigation:

- *Impact Assessment Project for Drought Early Warning in the Sahel*. This project, conducted from 1979 to 1986, was designed to detect and monitor drought and to use modeling to assess the crop conditions and yield potential in the countries within Africa's Sahel and Horn regions. It was led by the U.S. National Oceanic and Atmospheric Administration (NOAA) in support of a U.S. Agency for International Development (USAID) initiative. The study area extended over millions of hectares of largely inaccessible arid and semiarid land. Some of the many types of data that were interfaced in this project included remote sensing data from several different spacecraft, ground-based point data of varying reliability, a vegetation index proxy for crop growth and yield, interpolated soil properties, and integration of the data by use of a model. Of the six cases that the committee examined, this was the only international study. It involved coordinating data from 10 developing countries, which generally lacked the technology and resources to adequately support the NOAA/USAID efforts.
- *The National Acid Precipitation Assessment Program (NAPAP)*. This comprehensive research and assessment program was established by federal law in 1980 to, among other purposes, "evaluate the environmental, social, and economic effects of acid precipitation." It involved the cooperation of many different federal agencies and federally funded laboratories. The committee focused its review on the data management and interfacing activities related to the Aquatic Processes and Effects portion of the total program. The types of data collected included water chemistry and biology, wet deposition of acidic air pollution compounds, meteorology, hydrology, and episodic response of water bodies to acid deposition. These data were used to generate predictive models.
- *The H.J. Andrews Experimental Forest Long-Term Ecological Research Site*. Funded by the National Science Foundation, the Long-Term Ecological Research (LTER) Program consists of 18 independent sites in

the United States, of which the H.J. Andrews Experimental Forest in Oregon is one. The LTER Program studies are designed to carry out long-term research on diverse natural ecosystems. Their objectives are to study the following major features: pattern and control of primary production; spatial and temporal distribution of populations selected to represent trophic structure; pattern and control of organic matter accumulation in surface layers and sediments; pattern of inorganic inputs and movements of nutrients through soils, groundwater, and surface waters; and pattern and frequency of disturbance to the site. Research at the H.J. Andrews Experimental Forest, which was designated as a LTER site in 1980, has focused on several areas, including the disturbance regime, vegetation succession, long-term site productivity, and decomposition processes.

- *The Carbon Dioxide Information Analysis Center (CDIAC).* This center, located at the Oak Ridge National Laboratory (ORNL) and funded by the Department of Energy, provides high-quality data sets to the climate change research community. Its data management program exemplifies the kinds of data gathering, cleanup, documentation, and dissemination activities that are a necessary part of many data interfacing exercises. The types of data available include worldwide energy production statistics, population estimates, biological carbon dioxide sources and sinks, measured concentration of carbon dioxide, extensive related metadata, and numerous models.

- *The First ISLSCP (International Satellite Land Surface Climatology Project) Field Experiment (FIFE).* The long-term goal of ISLSCP is to improve our understanding of satellite measurements relating particularly to the fluxes of momentum, heat, water vapor, and carbon dioxide from land surfaces. The research goals of FIFE, which was conducted by NASA and several other agencies at a 3,400-hectare site in Kansas in 1987 and 1989, were to determine whether our understanding of biological processes on small geographic scales can be integrated over much larger scales to describe interactions appropriate for climate models, and to determine whether selected biological processes or associated states can be quantified over appropriate scales for climate models. The operational goals of FIFE included the simultaneous acquisition of satellite, atmospheric, and surface data; multiscale observations of biophysical parameters and processes controlling energy and mass exchange at the surface to determine how these are manifested in satellite radiometric data; and provision of integrated analyses through a central data system.

- *The California Cooperative Oceanic Fisheries Investigation (CalCOFI) Program.* This is an example of a long-term, broad-scale interdisciplinary research and monitoring program, the major goal of which has been to describe and understand the relationship between biological patterns and physical oceanographic/climate processes. The CalCOFI program has been under way since 1948 and is supported by the National Marine Fisheries Service, the California Department of Fish and Game, and the Scripps Institution of Oceanography. It exemplifies several scientific and organizational features that are important to the success of interfacing diverse data in interdisciplinary research.

The committee formed a separate subcommittee to evaluate each case study and established a list of evaluation criteria to help guide the fact finding. These criteria are presented in Appendix A. The subcommittees obtained briefings from the researchers and data managers, reviewed key background documents, and made site visits in all but the FIFE and Sub-Saharan Africa case studies. The subcommittees then reported back to the full committee both orally and in writing. The resulting case study reports were the product of the entire committee's deliberations.

ORGANIZATION OF THE REPORT

Chapters 2 through 7 present the results of the six case studies described above and an analysis of the major data interfacing issues that were identified through the case studies and related research. These chapters are all similarly structured, with sections containing the relevant background, the variables measured and the sources of data, the major data management and interfacing elements and issues, and the lessons learned. Some of the more complex case studies, notably the Impact Assessment Project for Drought Early Warning in the Sahel and the National Acid Precipitation Assessment Program, have a number of additional sections.

The final chapter provides a thorough overview of the issues and requirements related to the interfacing of diverse environmental data. It identifies the problem and its context, describes the barriers to effective data interfacing, and presents Ten Keys to Success for data interfacing activities. Supporting examples from the case studies are provided throughout.

REFERENCES

National Research Council (NRC). 1990. *Research Strategies for the U.S. Global Change Research Program.* National Academy Press, Washington, D.C.
National Research Council (NRC). 1992. *Combining Information: Statistical Issues and Opportunities for Research.* National Academy Press, Washington, D.C.
National Science and Technology Council (NSTC). 1994. *Our Changing Planet: The FY 1995 U.S. Global Change Research Program.* Government Printing Office, Washington, D.C.

2

Impact Assessment Project for Drought Early Warning in the Sahel

This case study was selected by the committee as a prime example of a project that used and attempted to integrate disparate data from many sources, including sequential satellite sensor data. The multinational area of sub-Saharan Africa (the Sahel) included in this case study extends across millions of hectares of fragile arid and semiarid land. For hundreds of years the nomadic human and animal populations of this vast region have been subjected to periodic drought and famine. As uncontrolled growth of these populations has increasingly denuded and degraded these fragile grazing lands, the frequency and devastating effects of famine have increased.

Rural populations in arid and semiarid regions of Africa are especially vulnerable to the effects of drought. Severe food shortages resulted from the African droughts of 1972-73 and 1983-84, particularly in the sub-Saharan region. Consequently, international and domestic agencies have increasingly emphasized the importance of a drought monitoring program.

As the United States sought to define meaningful ways to respond to the plight of and requests for technical assistance from the drought-affected sub-Saharan nations, one of the proposals that emerged was to use Earth observation (remote sensing) systems to provide regular sequential data about the crop growth conditions in the region (Salby et al., 1991). It had previously been demonstrated in many areas of the world that data from satellite sensors could be used to derive two kinds of information that would be useful in crop yield prediction—the first, a time-sequential inventory of quantity and type of clouds over the region (McDonald and Hall, 1980), and the second, a quick analysis of the extent and rate of growth of green vegetation over large areas. The major question posed by this proposal was whether data from Earth observation satellites could be combined with sparse data from other sources to design a system and methodology for credible crop modeling and yield predictions in this environment. Several U.S. government agencies therefore organized a pilot project to determine the feasibility of this approach.

The project, which was conducted from 1979 to 1986, had three objectives: (1) detection and monitoring of droughts, (2) crop modeling condition assessment, and (3) prediction of crop yield potential (LeComte, 1994). The crops that were monitored included cowpeas, maize, millet, peanuts, and sorghum. These objectives were interrelated in that water is the most critical limiting factor for crop growth in Africa's Sahel and Horn regions. Precipitation is usually the only source of water for growing crops in this area. In fact, average depth to groundwater in the region is so great that groundwater as a source for crops is not a serious alternative. The development of a capability to determine precipitation and its variability over the region in near real time could be expected to improve significantly the assessment of crop and yield potentials, both spatially and temporally.

The U.S. Agency for International Development provided most of the financial support, and the U.S. National Oceanic and Atmospheric Administration (NOAA) was the executing agency of the project (referred to as the "NOAA project" below). The Climate Impact Assessment Division (CIAD) of NOAA's Assessment and Information Services Center (AISC) was responsible for overall management. Appropriate agencies from each of the participating Sahelian and Horn countries provided valuable input to and collaboration with the project.

Major features of the NOAA project included the development of a plan of action or implementation plan and of a management structure and organization to implement the plan. Neither of these tasks was easy or clear-cut. In the formulation of each planning task, many assumptions had to be made, not the least of which was that adequate ground observation data could be obtained in a timely fashion. Another hope, if not an assumption, was that adequate in-country technical support for each of the participating sub-Saharan countries would be available with a minimum of constraints in the flow of essential information. Another feature, perhaps unique to this project, was the global perspective, which necessitated the integration of disparate global, national, and local data sources. This integration also required free flow of data and information across many international boundaries.

VARIABLES MEASURED AND SOURCES OF DATA

The methodology and variables used in the NOAA project are described in a technical report by LeComte et al. (1988). Ten African countries were included in the project, with each country generally having numerous crop growing areas. These different areas often had significant variations in environmental conditions. The project used groundwater measurements and satellite data for temperature and precipitation measurements. Rainfall was estimated for areas where data transmitted by the World Meteorological Organization's (WMO) Global Telecommunications System were not available or not reliable. In addition, 10-day rainfall data were supplied by the AGRHYMET (Agriculture-Hydrology-Meteorology Program) Center in Niamey, Niger. These measurements were summarized into an overall precipitation measurement for a given agroclimate region. Meteorologists used cloud imagery data from the NOAA polar-orbiting satellites and the European "Meteosat" geostationary weather satellite to supplement rainfall reports.

The meteorological satellite data were used throughout the project period to obtain cloud cover information, to monitor cloud movements, and to estimate precipitation patterns. Images over the entire sub-Saharan region were obtained every 3 to 6 hours from the Meteosat system. Unfortunately, this technique did not allow identification of local area convection.

The Meteosat data were compared with the mean outgoing long-wave radiation map published weekly by the Climate Analysis Center (LeComte et al., 1988). This was done by overlaying the two maps and drawing lines to identify discrepancies.

Historical data from the most recent 10 to 12 years were used to create baselines for normal rainfall and crop yield for each crop (millet and sorghum) within each of the agroclimate regions. There were regions or cropping areas for which these baselines could not be developed because credible data were not available. The historical data were obtained from the NOAA National Climatic Data Center, from site visits to the sub-Saharan region, or through correspondence with local meteorologists or other scientists. For historical precipitation data, the available dates and quality of data varied by country. For some countries, such as Sudan, these data were very poor. This variability resulted in the use of different years to establish baseline precipitation norms for each country. Another difficulty with precipitation data for the region was that the data were reported at different spatial scales for different areas.

The location of the Intertropical Discontinuity, its movement northward early in the season, and its subsequent retreat southward during the growing season were monitored daily. Determination of the Intertropical

Discontinuity's position depended primarily on station reports of dew point temperatures on the 1200 universal time coordinate surface analyses provided by the U.S. National Meteorological Center. The mean position of the Intertropical Discontinuity during the period from June to September has been found to correlate closely with cumulative rainfall and crop production in the sub-Saharan region (LeComte, 1994).

Advanced Very High Resolution Radiometer (AVHRR) data from the NOAA polar-orbiting satellites (NOAA, 1985) also were used to generate a Normalized Data Vegetation Index (NDVI) in the later stages of the project (Tarpley et al., 1984). The NDVI was calculated by taking the difference between reflectance measured by the visible band (0.58 to 0.68 micrometers (mm)) and the near-infrared band (0.72 to 1.10 mm) and dividing this difference by the sum of the reflectances of the two bands. The index was found to be correlated with the vigor and quantity of vegetative biomass (Tucker et al., 1985).

In general, there were serious discrepancies among data sources within a country. Subjective assessments and choices of data sources to use were made. Yield data especially varied greatly from country to country. Many interpolation algorithms were needed to integrate the wide variations in sparse ground observation data and data reported by different countries. Project planners had not anticipated that this set of problems related to yield estimates would be so serious. At least one subsequent review of the project criticized some of the methods used in the derivation of yield estimates (NRC, 1987). The committee found it difficult to make an adequate assessment of the methods used to arrive at yield estimates. The methods were not well documented, which led the committee to conclude that some aspects of yield estimate methods had been "ad hoc."

NOAA obtained rainfall estimates and gridded crop model output from the EarthSat Corporation. This information was used in confirming a planting date and various stages of crop production in order to mesh precipitation values with crop growth. The committee was informed through its briefings that the methodology used to generate these values was proprietary to the EarthSat Corporation and not available to anyone else, including the NOAA staff, though the planting dates did correspond to values estimated by using data from other sources.

In summary, an enormous volume of data was collected and summarized by the project, despite the generally spotty and sparse ground-based data. One of the greatest difficulties of the project was the lack of consistent and accurate data for the precipitation estimates and the yield models that were used. There were missing or unreliable data for such important parameters as precipitation, temperature, evaporation, date of planting of millet and sorghum, and areal extent of crops sown and harvested.

DATA MANAGEMENT AND INTERFACING

Before the initiation of the NOAA project, representatives from participating agencies met a number of times to formulate data management plans. These plans included procedures for relating relevant technical issues to data integration. Existing database technology and software, including some FORTRAN programs, the Statistical Analysis System (SAS) and Lotus 1-2-3, were used to process the data. No relational database or other procedures were developed specifically for the project. The committee was informed that data management requirements related to database interfacing probably did not significantly increase the overall cost.

A number of alterations were made to the data management plan during the project. Perhaps the most significant were made in response to the improvement in data acquisition, data handling and analysis technologies, and communications networking. Other changes were in response to difficulties encountered in obtaining desirable ground observations at the time they would have been most useful for real-time modeling of precipitation and yield predictions. If such a project were initiated today, the management and operations plan obviously would include much improved data acquisition, analysis, and communications systems. However, these new technologies would not eliminate the serious constraints imposed by the geographic and political boundaries. The remainder of this section identifies the most significant issues raised during the project regarding the management of data and database interfacing.

Timeliness of Data Acquisition

Time was a principal driver of the NOAA project. The very nature and objectives of the project meant that timeliness was critical for data acquisition (for both ground and remotely sensed observations), data analysis and interpretation, and distribution of yield prediction information to the decision makers and policymakers. Temporal uncertainties of precipitation and dates of planting contributed to the difficulties of acquiring optimal yield estimates. Also, a fundamental precept of the project was that weather conditions and crop yield predictions had to be made available to the users of that information as near to real time as possible.

Accessibility of Data

Accessibility of data varied with the type of data. Accessibility was generally determined by technical or political constraints. Some satellite images had to be distributed to users by mail or fax. It was necessary to perform extensive preprocessing of satellite data prior to their use in any models. This preprocessing was performed at the University of Missouri Cooperative Institute for Applied Meteorology.

Data distribution generally was exceedingly slow, in part because of the "primitive" technology available at the time, especially in the sub-Saharan countries. Each country provided some information, but it was not always in computer format. Also, the commercially obtained data for planting dates had a number of restrictions on use.

Data Quality and Verification

The NOAA project did not explicitly deal with the quality of data in written documentation. However, data quality is in question because a number of subjective judgments were made in weighing sources. This occurred with respect to both yield and precipitation, probably more so with the former. Some effort was made to check for errors in data recording, however.

The committee concludes that the measures of crop yield were not very accurate. The rank order of best to worst crop yields was probably correct, however. Periodic visits were made by personnel to the areas surveyed in an effort to verify and improve yield estimates.

Data Retention

Because the impetus for the NOAA project was the critical need for near-real-time crop yield predictions, it was designed to provide short-term data access, retrieval, and manipulation. Station rainfall and temperature data were loaded onto the AISC VAX 11/780 and stored for online access for 1 year. Cloud-cover data were entered based on visual interpretation of Meteosat visible and infrared bands of images sent to AISC.

Project personnel decided not to archive all the data because retaining such records was not considered important in the initial planning. Consequently, only some disks of data and related models are kept in the data center at the University of Missouri, and the data are not accessible electronically. The NOAA satellite data are archived at the National Climatic Data Center. Because different components of the project data reside in different locations, it is unclear how any subsequent crop modeling projects or activities might have access to and benefit from the data acquired during this project. The conclusion drawn by the committee is that little or no consideration was given to archiving the modeling data and supporting metadata for future use. It would be costly and an almost impossible challenge to recreate the data sets used in the models relating crop yield to precipitation.

Data Documentation

Because of the quasi-operational nature of the NOAA project, documentation of the data was not given a high priority. For instance, a set of Lotus spreadsheet files describing 10-day precipitation summaries was created with latitude, longitude, station name, and date. These are available from the National Climatic Data Center. However, the methodology (e.g., type of rain gauge and its locations) used to generate each point in the spreadsheet was not documented.

Another aspect that was not always well documented was the way in which missing data were handled. The committee was informed that there was a systematic effort to account for missing or uncertain data, but the specifics were not given. For example, it was noted that an analyst decided how heavily to weight the ground-station data with the satellite data, but an exact numerical basis for these weights was not in the reports provided to the committee.

Importance of the Crop Calendar

Familiarity with the crop calendar is of critical importance to the successful implementation of crop yield predictions in the arid and semi-arid sub-Saharan region. For instance, essential elements of the crop calendar for sorghum and millet in the environment in which the project was operating included time of planting, length of growing season, critical times during the growing season when the crop is most vulnerable to moisture stress caused by drought, and date of harvest. Each of these points in the crop calendar depends on when the rains come. In this region the beginning of the rainy season, if it begins at all, varies greatly, and the entire season is generally of short duration.

Knowing the crop calendar and following the precipitation events throughout the stages of growth and development of the grain crops are an essential part of conducting a credible crop yield prediction program, which can be modified as environmental and growth conditions change with advancing maturity of the crop.

Definition of Users of Crop Modeling Results

Two primary groups of users were served: (1) international agencies concerned with providing economic and food resources to the region under surveillance and (2) those national agencies involved in decision making related to production, internal and external trade, and processing and distribution of agricultural products. These users were clearly defined.

Climate assessments were made at regular intervals of the growing season (preplanting, planting, growth and development, maturity, and harvest). Crop yield estimates were delivered to Africa during the harvest season within days of obtaining the data in order to enhance maximum usage in those countries. The study was not designed to facilitate any significant access to data by users at a date much beyond harvest time or beyond the scope of the study.

Different user groups had different perceptions of the accuracy and detail that might be provided by the crop modeling project. Many ideas related to these perceptions had to be altered as the project progressed. The primary funding agency, USAID, found that the high cost of developing quantitative yield estimates in the work environment of the Sahel could not be justified, and it altered the work plan toward a much less labor-intensive approach to obtain more qualitative rather than quantitative yield prediction data.

Creation of New Algorithms and Data Management Procedures

To accommodate users' needs, new algorithms and data management procedures were developed. For example, coarse satellite cloud imagery was used to produce rainfall estimates in data-sparse areas. Reports were generated that were useful to analysts in the integration of these data with the available ground-based data. In addition, as discussed above, the NDVI was derived from the infrared and visible bands of the NOAA-9 AVHRR sensor to indicate growing conditions for 0.5° by 1° (latitude and longitude) grid cells across the region. Smoothed time series and regression models were used to integrate several components of the data. The data were typically accommodated in flat files.

Many of the algorithms used were not as well documented as one would like. For example, the use of a combination of ground-station data and satellite cloud-image data to derive precipitation estimates for a given locale was left to a climate assessment expert to interpret. The interpolation algorithms, which were different from country to country, were not accessible to the committee.

Accommodation of Users' Needs for Crop Yield Estimates

Officials of USAID appeared to be satisfied with the results of the yield predictions, and so the members of the committee were left to wonder why the NOAA project was terminated. The NRC (1987) report that was critical of the project yield estimates suggested that an alternative modeling procedure was in order. The approach suggested in that report would entail accounting for more of the underlying processes. However, the present committee concludes that it is unclear whether that approach would have been possible, given the constraints imposed in the project. One of the principal participants in the project thought that its high cost led to its termination. It is unlikely that an alternative modeling procedure would have alleviated these budgetary concerns.

Interfacing of Disparate Databases

As discussed above, the broad range of data used in the NOAA project included historical precipitation data used to establish normal ranges, satellite data to predict current rainfall, a mix of satellite-generated vegetation data, and historical and current yield data that came from several sources, including the countries surveyed. Yield estimates varied from country to country in quality and quantity. Important contributors to these discrepancies included the gross variations in and lack of supporting ground observation data, as well as the spotty nature of the precipitation.

The lack of uniform spatial and temporal scales created numerous problems. The density of meteorological stations in the area, the number of years of historical data, and the quantity and quality of meteorological data reported varied from country to country. A considerable amount of manual (e.g., noncomputerized) intervention was required to meld these data for continuity and format in a meaningful way.

There was a simple premise that crop yield could be adequately predicted from precipitation data. A major problem inherent in this approach is the quantitative assessment of the two factors—crop yield and precipitation—especially in the sub-Saharan environment with its severe sparsity of data.

Although in the original design of the project the importance of data interfacing was recognized, the extent and methods of such interfacing were not well documented. It therefore was difficult for the committee to assess the degree to which data integration was successfully accomplished. Nevertheless, even under optimal conditions, effective data interfacing in a project of multinational dimensions is extremely difficult.

Designers of the project sought to incorporate lessons learned from other crop yield experiments conducted over very large areas, especially from the federal interagency Large Area Crop Inventory Experiment (LACIE) and the Agriculture and Resources Inventory Surveys Through Aerospace Remote Sensing (AgRISTARS) program. In each of these experiments, estimation of yield was obtained from meteorological data, and areal extent of crop species was obtained from the analysis of sensor data from Landsat satellites (McDonald and Hall, 1978; NASA, 1983).

Institutional and Political Constraints

Any group that sets out to accomplish the kinds of objectives defined in the NOAA project immediately faces the dilemma of identifying institutions in participating countries that are equipped with facilities and personnel with the essential interests and skills for implementing the requisite tasks. Institutional constraints in many cases can be related to the fact that in some countries there is no clear-cut boundary or definition of the appropriate government agency to be assigned responsibility for collaborating in a project on crop modeling. Further, once an agency has been assigned responsibility, the identification of personnel with the special knowledge and skills essential to the project may prove to be difficult. These problems unfortunately were endemic throughout the sub-Saharan region.

Of course, institutional constraints are not unique to developing countries. Difficulties can arise within U.S. institutions that become involved in projects in international environments that are dramatically different from those in which these institutions normally operate. U.S. government agencies may be constrained by their mandates, organizational structure, modus operandi, and personnel with limited experience and skills required for successful international cooperation. Insufficient coordination and communication among participating agencies within the United States posed a problem in this project, as it has in many others.

Finally, political instability and subsistence-level economies increase the likelihood that essential data will be incomplete or inaccessible. For example, during the period of the project the war in Chad was a significant problem. However, even in those sub-Saharan nations not engaged in military conflicts, the generally low level of economic development and scientific and technical infrastructure magnified the problems associated with in situ data collection.

LESSONS LEARNED

Timeliness was essential for the realization of maximum utility of the results in the NOAA project. The objectives and scope were very broad and involved the interfacing of many disparate sources of data with wide variations in quality. As is frequently the case in such operational projects, formal data management was not given a very high priority. While data management was adequate to answer the immediate needs, little was done to organize the data for uses beyond the scope of the limited, short-range objectives. If data sets from these kinds of projects are to make any contribution to future efforts, provision must be built into the data management plan to ensure that the data sets, including essential metadata, will be preserved, archived, and made accessible to potential users.

The project involved the integration of several sources of precipitation data, and the "best professional judgment" was often used to combine the various values into a single number. All studies involve some degree of professional judgment. This project would have had greater value, however, if there had been more documentation describing the criteria agreed upon by the professionals who made these judgments. Such documentation is extremely important for anyone who wishes to use the data in subsequent studies.

A number of quality control procedures were exercised during various stages of the project. As used in this report, "quality control" refers to error correction, or to establishing and maintaining the validity and integrity of data. Unfortunately, little documentation of these procedures was recorded. The development of precipitation data summaries involved the integration of multiple sources, and quality control is essential to this activity. However, the yield data had a multitude of problems in deriving credible results. One inherent problem was the reliance on each participating country to provide accurate, timely ground observation data. Many of these countries did not have adequate technology, logistical support, and trained personnel to provide essential data. There were substantial efforts at quality control through collaboration with participating government agencies, often with a less than desired degree of satisfaction. Perhaps improved Earth observation satellites will mitigate this problem in the future.

It should be noted that an Earth-observing satellite may not last the lifetime of a project. In this project an effort was made to ensure that the values used from different satellite sensors were comparable. Any project of this magnitude that is so heavily dependent on sequential acquisition of data by Earth-orbiting satellites must include in its data management and operational plan the cross-calibration and verification of sensors from each successive spacecraft.

Portions of the project data reside on archived tapes. For those data to remain useful, the redundant backups should be kept, and the data must be transferred periodically to more current media. Directories that describe these data sets and their accessibility should be available.

Managers of research or assessment projects tend to believe that their projects could have accomplished much more if they had been adequately funded. This project was no exception, but its very nature and objectives gave it a special position among the case studies addressed by the committee. It was the only international study, and it dealt with a problem that is fundamental to numerous environmentally fragile areas of the world, many of which seem to be perennially on the brink of disaster from famine. The project held the promise of developing and testing a set of options and methodologies for monitoring and providing near-real-time information about growing conditions and yield predictions of food grains for the inhabitants of a very large area across sub-Saharan Africa. Unfortunately, inadequate funding made it impossible to provide the kind of crop monitoring and early warning system that was envisioned, although the project was successful in demonstrating the potential benefits of using advanced technologies for such applications.

REFERENCES

LeComte, D.M. 1994. The NOAA/NESDIS impact assessment project for drought early warning in the Sahel. In *Crop Modeling and Related Environmental Data: A Focus on Applications for Arid and Semiarid Regions in Developing Countries*. P.F. Uhlir and G.C. Carter, eds. CODATA, Paris.

LeComte, D.M., F.N. Kogan, C.A. Steinhorn, and L. Lambert. 1988. *Assessment of Crop Conditions in Africa*. NOAA Tech. Memo. NESDIS AISC 13. NOAA, Washington, D.C.

McDonald, R.B., and F.G. Hall. 1978. LACIE: An experiment in global crop forecasting. Pp. 17-48 in *Proceedings of the LACIE Symposium*. JSC-14551. NASA Johnson Space Center, Houston, Tex.

McDonald, R.B., and F.G. Hall. 1980. Global crop forecasting. *Science* 208: 670-679.

National Aeronautics and Space Administration (NASA). 1983. *Agriculture and Resources Inventory Surveys Through Aerospace Remote Sensing (AgRISTARS)*. Res. Rep. AP-J2-0393. NASA, Washington, D.C.

National Oceanic and Atmospheric Administration (NOAA). 1985. *Hydrologic and Land Science Applications of NOAA Polar-orbiting Satellite Data*. National Environmental Satellite, Data, and Information Service, Washington, D.C.

National Research Council (NRC). 1987. *Final Report: Panel on the National Oceanic and Atmospheric Administration Climate Impact Assessment Program for Africa*. Office of International Affairs. National Academy Press, Washington, D.C.

Salby, M.L., H.H. Hindon, K. Woodberry, and K. Tanaka. 1991. Analysis of global cloud energy from multiple satellites. *Bull. Am. Meteorol. Soc.* 72(4): 467-480.

Tarpley, J.D., S.R. Schneider, and R.L. Money. 1984. Global vegetation indices from the NOAA-7 meteorological satellite. *J. Clim. Appl. Meteorol.* 23: 491-494.

Tucker, C.J., J.R.G. Townsend, and T.E. Goff. 1985. African land-cover classification using satellite data. *Science* 227: 369-375.

3

The National Acid Precipitation Assessment Program

The Acid Precipitation Act of 1980 (Public Law 96-294) established a comprehensive 10-year research program to achieve the following purposes:

1. to identify the causes and sources of acid precipitation [defined as "wet or dry deposition from the atmosphere of acid chemical compounds"];
2. to evaluate the environmental, social, and economic effects of acid precipitation; and
3. based on the results of the research program established by this subtitle and to the extent consistent with existing law, to take action to the extent necessary and practicable (A) to limit or eliminate the identified emissions which are sources of acid precipitation, and (B) to remedy or otherwise ameliorate the harmful effects which may result from acid precipitation.

The terms of the statute established an Acid Precipitation Task Force, of which the Secretary of Agriculture, the Administrator of the Environmental Protection Agency (EPA), and the Administrator of the National Oceanic and Atmospheric Administration (NOAA) were designated as joint chairpersons. Membership of the task force also included one representative each from the Department of Interior, the Department of Health and Human Services, the Department of Commerce, the Department of Energy, the Department of State, the National Aeronautics and Space Administration, the Council on Environmental Quality, the National Science Foundation, and the Tennessee Valley Authority, in addition to the directors of the Argonne, the Brookhaven, the Oak Ridge, and the Pacific Northwest National Laboratories, as well as four additional members appointed by the President. The four National Laboratories were designated as a research management consortium to carry out a comprehensive and coordinated research plan, and the Administrator of NOAA was designated as the director of the overall research program.

In view of the complexity and extent of the 10-year National Acid Precipitation Assessment Program (NAPAP) study and the limitations of the committee's time to devote to this task, the committee decided to concentrate its review efforts on data management aspects of the Aquatic Processes and Effects portion of the total program. In addition to the documents specifically cited below, the committee reviewed a number of other publications relevant to this case study (NAPAP, 1990a,b,c,d,e,f,g; 1991a,b; Oversight Review Board of the NAPAP, 1991; Rubin, 1991; and Rubin et al., 1992.)

VARIABLES MEASURED AND SOURCES OF DATA FOR THE AQUATIC PROCESSES AND EFFECTS PORTION OF NAPAP

The highly diverse data and information needs for the Aquatic Processes and Effects part of the total study are summarized in the *National Acid Precipitation Assessment Plan* under two topics (Interagency Task Force on Acid Precipitation, 1982):

> first, the chemical alteration of water quality, including ground water, drinking water supplies, streams, and lakes; and secondly, the effects on the species and populations that make up biologically productive components of aquatic ecosystems. The information needs on water quality effects from acid precipitation concern regional trends, factors affecting watershed tolerances, the chemistry of metal mobilization, modeling, and related dose/response relationships for watersheds, lakes, and streams, and the risk associated with effects on drinking water.

Research components designed to obtain the needed information are presented under the following headings:

1. *Monitoring National and Regional Water.* "In addition to the water chemistry, factors to be documented should include: weather and acid deposition records; air trajectory data and the frequency of lightning (a natural nitrate production mechanism); soils, geology, and land use in the watershed and upwind areas; and watershed management trends that could affect the acid neutralizing and buffering capacity of the vegetation and soil."

2. *Determining Factors That Control Lake Susceptibility.* Analyses of "lake/environment relationships will indicate the relative importance of hydrogen ions from precipitation and dry deposition, relative proportions of nitrate and sulfate inputs, soil-chemical processes, predominant vegetation, and bottom sediment characteristics."

3. *Determining the Relative Contribution of Nitric and Sulfuric Acid Inputs.* "Studies will be undertaken to determine the relative contribution of nitrogen and sulfur from acid deposition to the productivity and/or acidification of aquatic ecosystems."

4. *Evaluating the Significance of Mobilization of Toxic Metals.* "Analyses will be made of the extent to which metal contamination in drinking water, food crops, and fish is due to acid deposition and subsequent leaching and mobilization of metals."

5. *Modeling Watershed Dose/Response Relationship.* "Attempts will be made to develop simple empirical models relating the readily measured chemical characteristics of lakes and streams to atmospheric deposition." "Relatively detailed simulation models of the acidification process and its effects will be developed and evaluated." "The goal of this research will be to have the most complete, quantitative long-term dose/response models evaluated fully and compared with the more empirical field relationships now in use."

6. *Studying Acidification of Drinking Water Sources.* "Analyses will be made of historical records and current data from public drinking water systems, whether using ground water or surface water reservoirs, to determine whether pH or potentially significant metal concentrations have changed during the past 10 to 30 years. Where acidification is found, the chemistry of water supply lines will be studied and estimates will be made of the possible impact on human and livestock populations."

7. *Monitoring Drinking Water and Evaluating Treatment Methods.* "Investigations will be made of how much effect chemical treatments, such as lime or other alkaline solutions, have on the acidity of surface water or ground water sources of drinking water. The possible short- and long-term usefulness of this ameliorative approach on human health will also be determined."

8. *Monitoring Regional Trends in Biological Effects.* "Scientists will seek to identify lakes and streams believed to have been affected by or apparently tolerant to acid deposition. Information on fish-eating birds of prey and furbearing mammals also will be sought."

9. *Studying Watershed Productivity.* "Measurements will be made of progressive changes in: (1) the chemistry of the open-water system and sediments; (2) the types and numbers of surface, subsurface, and bottom-dwelling insects, plants, animals and micro-organisms; and (3) terrestrial productivity (using predictive models when necessary). Efforts will be

made to establish correlations between the chemical properties of the water or lake sediments and the populations and reproductive success of the various organisms."

10. *Identifying Vulnerable Growth Stages.* "Field and laboratory experiments will be conducted with aquatic animals, plants, and micro-organisms to identify times of reproduction and stages of growth that coincide with episodes of strong acid inputs."

11. *Studying Metal Contamination of Fish.* "Analyses will be made of historical records, fish samples, and trophy fish to determine if concentrations of toxic metals in fish have changed over time."

12. *Analyzing Mitigation Strategies for Acidified Lakes.* "Experiments will include the application of various types of acid-neutralizing materials, such as powdered lime, rock limestone, and organic or inorganic materials that would bind or inactivate toxic metal ions."

The coordinating agency for this research was EPA. Other participating agencies included the Department of the Interior, the Department of Agriculture, and the Tennessee Valley Authority. Of the above research subject areas, numbers 1, 2, 3, 5, 6, 8, 9, and 10 were accorded priority 1, and the remainder (4, 7, 11, and 12) were priority 2.

As documented in the *National Acid Precipitation Assessment Plan*, the task force agreed to the following criteria for assigning research task priorities:

> *Priority 1*—Urgently needed research of the highest priority. Timely conduct of these research tasks is necessary to answer critical scientific questions concerning acid deposition. Each task investigates a crucial question where no or inadequate similar research is underway. The economic or social value of the potentially affected resources is high and the geographical area of investigation is highly sensitive to or heavily affected by acid deposition.
>
> *Priority 2*—Research that addresses an important information need but is less urgent than Priority 1. The phenomenon or geographical area to be investigated is believed to be moderately sensitive to acid deposition. The economic or social value of the affected resource is high.

Using these definitions, the coordinating agencies together with the participating agencies recommended, and the task force approved, the research priorities identified above.

EPA divided its Aquatic Processes and Effects portion of NAPAP into three major projects, entitled the National Surface Water Survey (NSWS), the Direct/Delayed Response Project (DDRP), and the Episodic Response Project (ERP). The committee's review of data interfacing activities focused primarily on the NSWS and DDRP. NSWS included these elements:

- A survey of water chemistry in a statistical sample of almost 3,000 lakes and streams representing a population of 28,000 lakes and 200,000 km of streams in acid-sensitive regions of the United States.
- Studies of watershed geochemical processes, deposition rates, fish toxicity, and temporal variation in lake and stream chemistry.
- Analysis of long-term chemical data, fishery records, and lake sediments to document historical changes in surface water chemistry.

DDRP's overall purpose was to characterize geographic regions by predicting the long-term response of watersheds and surface waters to acid deposition. The regions selected for study were chosen from regions with surface water that have low acid-neutralizing capacity and that exhibit a wide contrast, both in soil and watershed characteristics and in levels of acid deposition.

An additional biological assessment, the Episodic Response Project, was subsequently incorporated into the Aquatic Processes and Effects portion of NAPAP. This occurred well after the design of both NSWS and DDRP, when it became apparent that additional biological measurements would be necessary to achieve the NAPAP goal in this research area.

MAJOR CONSIDERATIONS IN EVALUATING THE DATA MANAGEMENT ACTIVITIES OF THE AQUATIC PROCESSES AND EFFECTS PORTION OF NAPAP

Users' Needs

Two major issues emerged under users' needs: identifying the users at the inception of the research and monitoring project, and understanding users' requirements.

The identification of the primary users was clear for the National Surface Water Survey (NSWS) because the whole project stemmed from a question asked by then EPA Administrator William Ruckelshaus regarding the status of acid-sensitive surface waters in the United States. Therefore, from the inception of the project, the principal user was clearly the administration of EPA at a very high level. Other portions of NAPAP included: terrestrial effects, effects on materials and cultural resources, visibility effects, economics, and atmospheric transport and deposition. Of course, the output of the Aquatic Processes and Effects portion of the program was also planned to be an input to the overall NAPAP integration synthesis.

For the lake survey portion of NSWS, research managers were particularly effective in maintaining good communication with the primary user identified at the beginning of the project. Although there was a notable exception, as discussed below, the designer/manager of the project focused only on identifying or answering the general question asked by EPA. Because some opposition to this approach was voiced, additional sharply defined questions were asked by the EPA Administrator: how many acid-sensitive lakes and streams are there, how sensitive are they, and where are they located? In contrast, some scientists wanted more longer-term process data and more detail on single ecosystems.

The committee concludes that the NSWS portion of NAPAP was successful in answering these questions and in providing useful data and information to the primary user. The NSWS director avoided vague mandates and tried to be as specific as possible in defining goals. In addition, desirable interactions between scientists and policymakers were maintained throughout most of the program. One great strength was the continuity of scientific project leadership throughout the program—the project leaders knew the program goals and stayed with them. Perhaps even more important was the relatively stable, high-level support for NSWS within EPA.

The above comments concerning the continuity of project leadership and the stable high-level EPA support apply equally well to the follow-on study on critical watersheds, the DDRP. The primary user for this study was the relevant program office in EPA. The EPA staff assured the committee that the question that DDRP was trying to answer came from the EPA Administrator Ruckelshaus and his concern for what the future would bring; namely, what types of systems were vulnerable, where did they occur, and how would they respond under various emission control scenarios (including a no-action option)?

These DDRP questions were translated into watershed-level and process studies by the EPA research and design team. This was a very complex set of objectives. The success of the DDRP in answering these questions was less obvious to the committee than in the case of the NSWS. The users here included both some scientists working on the project and the policy- or decision-making staff of EPA. Although there may have been some conflicts between EPA policymakers and the scientists, it seems clear that the overall study was designed so that the primary data users would be policymakers at EPA.

Study Design

There are two key areas related to study design: conceptual models and methodological considerations. In general, these two areas cannot be independent and must be mutually supportive.

In the initial NSWS sampling, no conceptual model for the ecosystem was apparent. The choice to measure acid neutralizing capacity (ANC) was based on a model of how lake chemistry works, and expert groups were used to determine variables that would be measured. Although major cations and anions were analyzed, ANC turned out to be the key variable not only in the survey of lake sensitivity, but also of the overall NAPAP, for several reasons. First, although ANC is an aquatic ecosystem variable, it integrates conditions in the watershed and is itself a function of various terrestrial processes (including processes of the soils, biota, and parent bedrock). Second, focusing on a single variable aided in briefing policymakers because they could understand and use the data relatively easily and were therefore likely to continue their support of the program. Third, ANC is now universally recognized as the key variable indicating acid sensitivity for aquatic ecosystems. It is thus to EPA's credit that it recognized early the importance of ANC.

The success enjoyed with ANC may not be easily translated into lessons learned for other complex programs, however. Multiple pollutants and impacts from various pathways may preclude an easy focus on a single ecological parameter.

Identification of a planned statistical analysis seemed to be the first priority in the NSWS project's experimental design. This approach was successful because the questions asked by the user and the background knowledge of the design team meshed well. The resultant probability sampling for NSWS was considered to be the most desirable approach by the policymakers at EPA, as well as by the committee. This approach facilitated follow-on programs, such as EPA's Environmental Monitoring and Assessment Program, and helped to maintain political support for the NSWS program. Two aspects of the NSWS project design process are worthy of note: (1) the data users' views were sought early in the design of the study, and (2) policymakers were strongly influencing the direction for this program and future EPA programs.

The use of conceptual models in the design of Aquatic Processes and Effects watershed-intensive studies was much more evident. Here the questions were more complex, and the designers took a much longer time to review existing conceptual models and develop new ones. Reliance on a dispersed stochastic sampling design was not feasible, and the designers relied much more heavily on the use of deterministic and conceptual models, both in the design and in the interpretation of the collected data. In addition, expert groups were used to help select where and what actual measuring points needed to be sampled. It seems to the committee that many of these experts relied heavily on historical data sets. Because it was not possible to monitor every water body, models were required that would permit statements to be made about many water bodies on the basis of limited data.

In the watershed parts of the Aquatic Processes and Effects studies, EPA researchers did have to integrate multimedia data, including data on atmospheric deposition, watershed, soils, and surface water chemistry. Because of this need, they also were more dependent on a good conceptual model, not only in the planning stages but, more importantly, in their final assessments, which made projections and predictions on a regional basis. The committee concludes that the technical aspects of these parts of the project were performed successfully.

There was a fortuitous aspect to the watershed parts of the studies that was key to the successful turnaround and interpretation of data. Sulfate concentrations did not vary appreciably with time (seasonally), and so extensive spatial data could be used in assessments without expensive temporal characterization. It was to EPA's credit to realize (and document) this circumstance early on and to take advantage of it in both the conceptual and the practical design of the project.

Unfortunately, the biological design portion of the Aquatic Processes and Effects part of NAPAP had more difficulty from a conceptual and budgetary standpoint than the physical/chemical sampling and measurement tasks. For example, some scientists who helped plan the biological sampling efforts believed that too much emphasis was placed on the physical/chemical parameters in the initial design and not enough on the biological needs. In particular, scant attention was paid to the selection of chemical parameters that were most important for understanding biological impact, and funds were limited for implementing biological measurements. While NAPAP had excellent experimental data for fish response to acidification (a well-focused impact) with insights on interactions with calcium and pH, there was an insufficient effort to collect new biological data from the field that were integrated with concurrent measures of physical/chemical parameters deemed important in the earlier aspects of the aquatic studies. Apparently, some adjustments for this deficiency were made midway through the biological design portion.

The methodological considerations reflected some of the strengths and weaknesses associated with the conceptual framework for the three parts of the Aquatic Processes and Effects portion of NAPAP reviewed by the committee. For the NSWS, the method of ANC determination in terms of sample collection, preservation, and laboratory analysis had to be developed and tested; the method is now consistent, accurate, precise, and regularly applicable to a wide range of aquatic ecosystems and habitats. Thus, from a procedural view, the focus on ANC in the NSWS was a considerable asset in terms of managing data of known quality and for later data integration activities.

In contrast, the methods for biological assessment frequently were a liability to the success of the program. The hydrogen ion and metal concentrations (especially aluminum) of surface waters are more relevant to determining acid rain impacts to fisheries. The pH is difficult to measure in either the field or the laboratory for water of low ionic strength (typical of acid-sensitive ecosystems). Methods of measuring metals have to take into account chemical speciation and dissolved versus particulate fractions; both the instrumentation and the methods for metal determinations are more complex and costly than for ANC. The most significant methodological problem, however, is in the sampling of fish populations. A variety of different methods, including the use of seines, nets, electroshocking gear, and poisons (e.g., rotenone), are typically employed for fisheries work. The efficiency of each method may differ with fish species, age class, habitat, and the field personnel; further, there is no consistent approach or regular coordination from study to study in the use of these various methods. Given NAPAP's dependence on existing fish data from state management agencies, coupled with these methodological problems (and the lack of good methods documentation), it is not surprising that data management and integration were problematic.

The effect of methods on the watershed part of the NSWS might be viewed as being somewhere between these two extremes. In essence, both a hydrological and a chemical budget (e.g., inputs, transformations, outputs) had to be measured for a given subcatchment for this part of the study. Although many more variables were measured than for the lake survey component of NSWS, both field and laboratory methods were more standardized than they were for biological measurements, or at least could be agreed upon; also, it appeared that some complex processes, such as water movement or ion exchange in soils that are difficult to measure, could be simplified for assessment purposes.

DATA MANAGEMENT AND INTERFACING IN THE AQUATIC PROCESSES AND EFFECTS PORTION OF NAPAP

In the following discussion, no attempt is made to treat systematically the specifics of all the various types of data collected in the Aquatic Processes and Effects program, or the specifics of the relationships between or among the data types. Similarly, the specifics of the data management system, which was an ad hoc system assembled by the contractor to provide verified data summaries in forms most available and helpful to the researchers, are not described in detail. Instead, the committee provides a summary of its most important observations dealing with problems related to data management and interfacing.

The committee identified many data management and integration issues associated with this complex research project. Among the most significant related to the actual procedures of how and by whom the data were stored and manipulated. The data management was subcontracted at a different site. There the data were verified and placed in a data management system and then distributed to the researchers.

Several issues arose with this arrangement. First, there was conflict over "ownership" of the data and timely return of the data to the scientific and technical team in EPA, most members of which were located at a different site and institution than the data management team. Briefings by both teams provided the committee with a number of insights that could help in the design and management of future interdisciplinary projects. The scientific and technical team left the impression with the data management team that the latter team's primary role was to provide a workable database from which the scientific and technical team and its subcontractors could do the actual data analysis and interpretation. However, data management team members considered themselves scientists as well as data managers and wanted the opportunity to interpret the data also. At stake was the issue of scientific recognition and credibility. According to the scientific and technical team, this issue had been equitably resolved by a series of early agreements governing data use and publication rights. Nevertheless, it was apparent that this matter was not viewed in the same light by the data management team even several years after the program was over. Without having the opportunity to work with the data, the data managers considered themselves inhibited in their ability to write and produce peer-reviewed publications describing their work.

Second, there were inordinately long delays in acquiring data from the chemical analysis laboratories. Such delays are a chronic problem in large-scale environmental sampling efforts that depend heavily on chemical analysis. EPA was able to correct this problem toward the end of the project by using a management tracking system and having much of the data verification done by the data management team. Long delays complicated matters because analytical problems could continue unchecked for quite some time, thus compromising data quality. EPA researchers eventually automated their audit program (e.g., checking variances and means plus outliers) and could notify analytical laboratories quickly to correct problems.

Third, the spatial location for the sampling data (deposition, soils, watershed, water chemistry, fishes) that were to be integrated for assessments illustrates another concern. In NAPAP, it appears that multimedia data were collected in relative physical proximity to each other, thus facilitating integration for a site or ecosystem. Designers of new research and monitoring programs such as EPA's Environmental Monitoring and

Assessment Program should be aware that if different media are randomly sampled independently of each other (different laboratories or institutions may have different media and may randomly select their own independent sites), data analysis and integration across media may be complicated.

Despite the difficulties summarized above, the database management system for the NSWS appears to have been well planned. EPA especially wanted it to be of a known quality. Both NSWS and DDRP had extensive quality control. The quality control measures included traditional quality control methods and appeared to be applied also to the metadata collected. The Quality Assurance Program was peer-reviewed, the database was verified (poor data were flagged), quality assurance (QA) data and metadata were included, and a database dictionary was put together. A number of peer-reviewed articles were published from NSWS, and the data sets were made readily available on diskette to a variety of other users, who have made numerous requests.

The database management team appears to have grappled with the question of how much QA is enough. Although they indicated that dedicating about 10 percent of fiscal resources may have been reasonable, in reality they may have spent about 20 to 30 percent of their funding on QA in some cases, based on some committee members' direct experience with NAPAP protocols. The database management team indicated that good QA early on is important, and they seemed to have pioneered new ground in this area, by solving the problem of high nitrogen in field blanks (from washing filters with nitric acid!) and developing natural audit samples that were much more useful for problems in limits of detection. In addition, by maintaining flexibility in QA, they were able to identify problems as they arose and to deal with them effectively.

It was more difficult to evaluate data management for DDRP because the final product was not yet available at the time that the committee conducted its case study. In general, the DDRP database is being developed along the same lines as the NSWS model, although with more complicated statistical analyses and including a data dictionary. The same contractor was used for managing both DDRP and NSWS data. The advantage was that the DDRP had the benefit of experience from NSWS, which provided good continuity in the program. The disadvantage was that logistics were complicated. Sites were in the East, project management was at Corvallis, Oregon, soil analysis was conducted by contract laboratories, and the data managers were at the Oak Ridge National Laboratory in Tennessee. Under this arrangement, communications were extremely difficult. A major decision for future studies was to conduct data management "in-house" to facilitate logistics and communications if the staff and hardware were available.

The data manager whom the committee interviewed provided some practical views on aspects of data management within the aquatic parts of NAPAP. He emphasized that data management should be about 10 percent of the total project budget, and this guideline apparently has been followed. His comparison of NSWS and DDRP was insightful: raw versus synthetic data; differences in management style; homogeneous, focused data sets versus complex multimedia data; and straightforward reporting of results (ANC emphasis) versus complicated analyses and predictions. Also, for DDRP, many data had to be carefully evaluated (e.g., geology and soils) because the format or units used were not consistent across states. For various reasons, including a mid-course expansion of the project, it took 3 years longer to get all of the data together than originally thought.

These differences highlight several issues relevant to the successful interfacing of data. The NSWS data sets required less documentation of metadata and were available relatively quickly to policymakers, agency administrators, and scientists both inside and outside of the agency. Because the NSWS data were more descriptive in nature, focused on a single medium, and required less preliminary processing, there seemed to be fewer organizational barriers to sharing and integrating data. With fewer variables in NSWS, it was easier to agree on and use standard formats and data conventions, which resulted in fewer data incompatibilities. Also, there was only a limited temporal component in NSWS, and all data were collected on the same spatial scale within a consistently used sampling design, a situation that facilitated data interfacing within various statistical analyses.

The multimedia database in DDRP was still unavailable at the time of this writing because of the greater degree of complexity in data integration. Scientists in a wider range of disciplines had different attitudes and approaches to sharing data, providing metadata, and using various types of software and hardware. Also, there was more uncertainty in the nature of the research, and so early results, especially experimental outcomes, resulted in shifting data requirements. For example, some pieces of field equipment did not work as well as expected or needed.

Perhaps the most significant difference between DDRP and NSWS was the requirement in DDRP of relatively complex models for data integration. Two such models were the Electric Power Research Institute's Integrated Lake-Watershed Acidification Study (ILWAS) model (Gherini et al., 1985) and the Model of Acidification of Groundwater in Catchments (MAGIC) from the University of Virginia (Cosby et al., 1984). Both models are driven by precipitation rates and rainfall chemistry (inputs) and are capable of predicting future (50 to 200 years) rates of surface water acidification (output) in terms of changes in pH and ANC. ILWAS, considered to be more realistic by some researchers, is complex and requires concurrent hydrologic and chemical process data on similar scales, especially in regard to soils that may be highly heterogeneous. In contrast, MAGIC is based on more simplified assumptions about hydrology and data for averaged soil parameters; thus, data integration is an easier task, but this benefit comes at the expense of less realism about the ecosystem.

Another degree of complexity in data interfacing develops when atmospheric fate and transport models are coupled with watershed models such as these. Clearly, coupled models are needed to produce the final product (future scenarios of acidification), and as such they determine the requirements of these models in terms of data preparation, quality control, data compatibility, and data management.

With regard to the biological assessment portion of the Aquatic Processes and Effects part of NAPAP, the committee found it difficult to understand why many aspects of that research, including data management in particular, deteriorated over time. Although NAPAP managers improved their efforts based on their experiences with the program, they were unsuccessful in obtaining sufficient funds to implement fully the biological component as originally proposed. Thus, the biological assessments portion, especially near the end of the program, seemed to have experienced the most problems: the database was not maintained nor accessible; no consideration was given to distributing or archiving the data; and there was poor planning (or no planning) early on concerning data needs, compatibility, and integration. It is important to note that policymakers were convinced from the total integrated results of the studies, including those on biological indicators, surface water chemistry, and paleolimnological studies (not described here), that adverse impacts on water bodies are occurring due to anthropogenic causes and that further research is indicated.

Although important difficulties with the aquatic parts of NAPAP have been identified, most of the individual components seemed to complement the others, and the database management system appeared to facilitate necessary interactions, by easing the exchange and application of data collected in one part of the program to another. Specifically, the tiered approach was thought to be successful: NSWS was an extensive survey of a more focused nature (fewer parameters and questions addressed), DDRP was regional and predictive in nature and integrated multimedia data, and the ERP focused on process-related research at fewer selective sites, but did integrate physical and biotic databases. This tiered approach was a deliberate part of the design of the project. Consideration was given to different spatial and temporal scales, and there was a balance between monitoring, assessment, and research to address process questions within a freshwater ecosystem perspective. This foresight appeared to be based on effective planning done by strong project management.

Finally, the committee was informed by the executive director of NAPAP, Derek Winstanley, that the future of the databases generated for the Aquatic Processes and Effects portion of NAPAP (and perhaps all of NAPAP) is uncertain. These data will be crucial to developing other programs, such as the Environmental Monitoring and Assessment Program, but no consideration was given to long-term maintenance of the data in the original goals, despite the long-term support of NAPAP. This situation should not be allowed to occur in future environmental monitoring programs.

LESSONS LEARNED

The lessons learned are organized so as to emphasize that data integration in interdisciplinary studies cannot be viewed as a separate and distinct entity. On the contrary, data integration is inextricably linked to program planning and objectives, all aspects of sampling and analysis, and the various methods and procedures employed in the analysis and interpretation of the resulting data. Accordingly, even though the principal topic of this study is data integration, the committee found it necessary to expand its review somewhat, in order to place data integration in the proper context.

This section summarizes the major lessons learned by the committee in its review of this interdisciplinary, multimedia, and exceedingly complex research program. It should be emphasized that NAPAP was a policy-oriented activity, and the potential difficulty in transferring the lessons from this to other types of studies is recognized. The major deficiencies in data management and integration were related principally to inadequacies in the program planning component.

Program Planning

An organization and management structure setting forth roles, responsibilities, and authorities of all cooperating agencies should be established at the outset. This should include the designation of a program director with appropriate authority over all participants.

A detailed experimental protocol should be developed and approved for the total project. This protocol should identify research needs, priorities, milestones, and a phased or tiered approach to completing the entire program. All foreseeable interdisciplinary research requirements and supporting data management provisions should be addressed at this time.

Multiyear commitments for required levels of resource support should be made and vigorously supported by each participating agency.

Methods, techniques, and procedures should be established in the planning process for moving the program forward successfully. Requirements for databases and data management to include acceptable data and metadata characteristics, formats, data ownership, and accessibility to users should be established. Such planning requires early consideration of how the data produced by each cooperating agency will be meshed to provide an integrated assessment of causes and effects of acid precipitation.

Provisions should be made for a periodic planning review process (e.g., every 6 months) to assess progress of the study and make any indicated mid-course corrections in the total experimental plan as well as in the technical work plans of the participating agencies.

Program Implementation

During program implementation, there should be timely exchange among all cooperating agencies of all research results and technical reports and presentations. This should involve frequent scientist-to-scientist technical exchanges, with special attention given to interfaces or boundaries among the research projects being carried out by the various cooperating agencies.

There should be periodic (e.g., yearly) internal and external quality assurance audits of all aspects of the total program. In addition, periodic external peer review of the total program implementation should be conducted.

Comprehensive information meetings involving all cooperating agencies should be conducted on at least an annual basis, with appropriate emphasis given to the interdisciplinary research issues and related data management and integration activities.

Program Completion, Evaluation, and Future Activities

Continuous feedback should be obtained from users of the resulting databases with regard to their accessibility, utility, and any problems encountered. Appropriate changes should be made as required.

Databases should be maintained in readily available form and updated as necessary. Resources required to accomplish this maintenance should be identified and obtained. Based on this case study, allocating at least 10 percent of the total program budget for data management would not be unreasonable.

Depending on research objectives, the level of effort and the scale of the required biological data should be carefully matched with those of the required geophysical and geochemical data during the planning phase of the total program.

Data managers and scientists should be located at the same site, if at all possible, to facilitate effective interaction and cooperative team efforts.

Provision should be made for periodic (e.g., once every 5 years, or more often) information meetings or symposia to review the scientific state of the art. Following each information meeting, a proceedings volume should be prepared, including a description of current interdisciplinary research needs and related data management priorities to support those needs.

REFERENCES

Cosby, B.J., R.F. Wright, G.M. Hornberger, and J.N. Galloway. 1984. *Model of Acidification of Groundwater in Catchments.* Draft Users Manual. EPA/NCSU Acid Precipitation Program, North Carolina State University, Raleigh, N.C. 246 pp.

Gherini, S.A., L. Mok, R.J. Hudson, G.F. Davis, C.W. Chen, and R.A. Goldstein. 1985. The ILWAS model: Formulation and application. *Water, Air, Soil Pollut.* 26: 425-459.

Interagency Task Force on Acid Precipitation. 1982. *National Acid Precipitation Assessment Plan.* Washington, D.C.

National Acid Precipitation Assessment Program (NAPAP). 1990a. *Current Status of Surface Water Acid-Base Chemistry.* Acidic Deposition: State of Science and Technology, Rep. 9. NAPAP Office of the Director, Washington, D.C.

_____. 1990b. *Watershed and Lake Processes Affecting Surface Water Acid-Base Chemistry.* Acidic Deposition: State of Science and Technology, Rep. 10. NAPAP Office of the Director, Washington, D.C.

_____. 1990c. *Historical Changes in Surface Water Acid-Base Chemistry in Response to Acidic Deposition.* Acidic Deposition: State of Science and Technology, Rep. 11. NAPAP Office of the Director, Washington, D.C.

_____. 1990d. *Episodic Acidification of Surface Waters Due to Acidic Deposition.* Acidic Deposition: State of Science and Technology, Rep. 12. NAPAP Office of the Director, Washington, D.C.

_____. 1990e. *Biological Effects of Changes in Surface Water Acid-Base Chemistry.* Acidic Deposition: State of Science and Technology, Rep. 13. NAPAP Office of the Director, Washington, D.C.

_____. 1990f. *Methods for Projecting Future Changes in Surface Water Acid-Base Chemistry.* Acidic Deposition: State of Science and Technology, Rep. 14. NAPAP Office of the Director, Washington, D.C.

_____. 1990g. *Liming Acidic Surface Waters.* Acidic Deposition: State of Science and Technology, Rep. 15. NAPAP Office of the Director, Washington, D.C.

_____. 1991a. *Mission, Goals, and Program Plan Post 1990.* Public Review Draft. NAPAP Office of the Director, Washington, D.C.

_____. 1991b. *The U.S. National Acid Precipitation Assessment Program 1990 Integrated Assessment Report.* NAPAP Office of the Director, Washington, D.C.

Oversight Review Board of the National Acid Precipitation Assessment Program. 1991. *The Experience and Legacy of NAPAP.* NAPAP Office of the Director, Washington, D.C.

Rubin, E.S. 1991. Benefit-cost implications of acid rain controls: An evaluation of the NAPAP integrated assessment. *J. Air Waste Manage. Assoc.* 41(7): 914-921.

Rubin, E.S., L.B. Lave, and M.G. Morgan. 1992. Keeping climate research relevant. *Issues Sci. Technol.* VIII(2): 47-55.

4

The H.J. Andrews Experimental Forest Long-Term Ecological Research Site

The H.J. Andrews Experimental Forest was established by the U.S. Forest Service in 1948. Located in the rugged Cascade mountain range of Oregon, the 6,400-ha preserve was covered with virgin forest in the late 1940s. Since then, approximately one-third has been manipulated through logging or research plantations. Old-growth forest stands with trees over 400 years old cover about 40 percent of the area, with mature stands covering another 20 percent. Rapidly flowing mountain streams are the primary type of aquatic system.

The Andrews Forest was designated as a research site under the National Science Foundation's (NSF) Long-Term Ecological Research (LTER) Program in 1980. The Andrews site is one of 18 such sites in the United States. The goals and objectives of the LTER Program are given in Franklin et al. (1990), and the research programs and core data sets of the Andrews site are summarized in McKee et al. (1987) and Michener et al. (1990), respectively.

The LTER Program studies are designed to carry out long-term (decades to two centuries) ecological research on natural ecosystems in the United States. Their basic objectives are to study:

1. Pattern and control of primary production;
2. Spatial and temporal distribution of populations selected to represent trophic structure;
3. Pattern and control of organic matter accumulation in surface layers and sediments;

4. Pattern of inorganic inputs and movement of nutrients through soils, groundwater, and surface waters; and

5. Pattern and frequency of disturbance to the site.

Research at the Andrews site has focused on several areas, including the disturbance regime, vegetation succession, long-term site productivity, and decomposition processes. The commitment to long-term studies is evident in these areas. A good example of this is a log decomposition study, which will determine the effects of log size and quality and of the site environment on the pattern and rate of decomposition and nutrient release. In the largest and longest decomposition experiment, more than 500 logs of four species were placed at six old-growth forest sites. That study is designed to track samples over a 200-year period (Harmon, 1992).

The scientific scope of the committee's case study is limited to the interdisciplinary observational and experimental studies at the LTER Andrews site, although it also reviews the data management and institutional relationships of the Andrews site to the other LTER sites and to NSF.

The research at the Andrews site has certain key similarities to the committee's other case studies. It was intended to sample and study interdisciplinary problems and issues. For example, it includes coordinated studies in air, soil, water, and various forms of biota. The research was well under way, having begun several decades before it was officially designated as an LTER site in 1980. The data collected could be expected to be useful in global change studies and in other types of long-term environmental monitoring efforts. And, finally, a variety of investigators worked on the same general study area.

The committee was primarily interested in the data management activities of the Andrews site. These activities are managed by the Quantitative Science Group under the auspices of Oregon State University and the U.S. Forest Service. There appears to be little distinction between whether an individual works for the university or the Forest Service. This arrangement seems to work well for a number of reasons. There has been a long and close working relationship between the U.S. Forest Service Research Laboratory and the Forest Science Department at Oregon State University's College of Forestry, as well as other departments. The proximity of the two buildings housing the respective scientists also promotes good collaboration. In fact, several university researchers and staff have their offices in the Forest Service Laboratory building. In addition, there has been a history of successful preparation of joint research proposals between university and Forest Service staff. This, plus traditional attributes of working at a university, such as joint appointments and cooperating faculty appointments, has helped to maintain an effective working relationship.

VARIABLES MEASURED AND SOURCES OF DATA

Table 4.1 lists and describes all of the data sets collected at the Andrews site.

TABLE 4.1   H.J. Andrews Experimental LTER Site Data Sets

Data Set	Description
Dendrometer measurements in permanent reference stands	Provides an accurate estimate of volume and height for individual trees.
Respiration patterns in logs	Examines the seasonal and successional patterns of respiration losses for four dominant softwood species.
Coarse woody debris density and nutrient content	Describes the external characteristics of coarse woody debris in various decay classes and measures density and nutrient content.
Stream cross-sectional profiles	Monitors changes in channel geometry in response to storms and movement of large organic debris in a range of stream sizes.
Watershed streamflow summaries	Evaluates long-term changes in hydrology associated with various management treatments; provides baseline data for affiliated nutrient, water chemistry, and sediment transport studies; and characterizes the hydrologic regime of old-growth forests at different elevations.
H.J. Andrews watersheds 1, 2, and 3 and miscellaneous suspended sediment samples	Quantifies long-term effects of two intensities of timber harvest on sediment delivery at seasonal and yearly time scales.
Post-logging community structure and biomass accumulation	Patterns plant succession and biomass accumulation following clear-cut logging of an old-growth Douglas fir/western hemlock forest.

TABLE 4.1 Continued

Data Set	Description
Plant biomass dynamics following logging and burning	Documents patterns of plant succession after clear-cut logging and slash burning on two experimental watersheds.
Tagged log inventory	Tags and numbers woody debris and describes the following characteristics: longitudinal position, geomorphic location, log dimensions, decay class, origin, moss cover, root wad, and channel angle.
Population studies of rainbow and cutthroat trout	Assesses fish population and habitat structure in streams (150 to 300 m in length) and basins (greater than 40 km in length).
Watershed 1 and 3 plant succession data, 1962 to 1977	Documents patterns of plant succession after clear-cut logging and slash burning on two experimental watersheds.
Tree permanent plots of the Pacific Northwest	Examines rates of succession and measures mortality and growth in representative forest types in Pacific Northwest.
Stream-upland wood decay experiment	Examines and contrasts the decay of small logs in a stream channel to that on an upland site; examines the movement of small logs in a third-to-fourth-order stream.
Reference stand litterfall study	Determines seasonal and annual rates of litterfall samples at six permanent plots picked to represent a range of habitats and elevations.
Structure and composition of riparian vegetation	Measures biomass of riparian vegetation strata, characterizes phenology of leaf-out and leaf fall, and determines the spatial distribution of foliar biomass, and timing and amount of annual foliar inputs into steams.

continues

TABLE 4.1 Continued

Data Set	Description
Rainwater samples: long-term precipitation chemistry patterns	Precipitation chemistry sampled at a low-elevation site and analyzed for pH, alkalinity, conductivity, total P, ortho-P, total N, NO_3-N, suspended sediment, Si, Na, K, Ca, Mg, SO_4-S, and Cl.
Watershed grab samples: long-term stream chemistry patterns	Describes long-term patterns of nutrient output from: a first-order, old-growth watershed; a first-order watershed after clear-cutting; a second-order old-growth watershed; a second order watershed logged and burned in 1966; and a third-order old-growth watershed. Provides baseline environmental monitoring data for studying nutrient availability for stream organisms, and recovery patterns of disturbed watersheds.
Nicotinamide adenine dinucleotide phosphate (NADP) precipitation chemistry	Measures precipitation samples collected weekly for pH and conductivity on site. Samples are then mailed to a Central Chemical Laboratory and analyzed for Ca, Mg, K, Na, NH_4, NO_3, SO_4, PO_4, pH, and conductivity.
Watershed proportional samples: long-term stream chemistry patterns	Stream chemistry sampled to characterize the timing and amount of elemental losses in undisturbed conditions, and to determine the effects of logging on rates of nutrient release.
Primary meteorological station at headquarters	Provides climatic summaries and documentation for the primary meteorological station at H.J. Andrews, 1972 to present.
Climate station at watershed 2	Continuously records precipitation, relative humidity, and air temperature.

TABLE 4.1 Continued

Data Set	Description
High-elevation meteorological station	Takes measurements of air and soil temperature, soil moisture equivalency in both clear-cut and shelterwood; and solar radiation, precipitation, and wind speed and duration in the clear-cut.
Rain gauge network	Provides baseline information on variation in precipitation across a wide range of site conditions.
Air, soil, and stream temperature in various habitats in and around the Andrews Forest	Continuously monitors air, soil, and stream temperature at selected habitats.
Snow survey	Provides a baseline for characterizing variation in snow depth, moisture, and duration in the western Cascades for hydrologic modeling and to distinguish the differences in the microclimates of dominant plant communities.
Plant component biomass equations and data for the Pacific Northwest	Contains data on biomass, leaf area, and sometimes other measurements of plants collected in the Pacific Northwest.

Source: Michener et al. (1990).

DATA MANAGEMENT AND INTERFACING

The LTER concept is to have a network of intensively studied sites around the United States in various ecosystems, all measuring similar parameters and studying similar ecosystem processes. Long-term monitoring of selected environmental parameters is also a major objective (see Institute of Ecology, 1981). The LTER Program, while meeting some of these goals, has developed more along a principal investigator driven agenda with all the diversity of research that implies. A major reason is that NSF has not posed any specific research questions and only very broad goals for the various LTER sites. Therefore, the individual investigators have considerable autonomy in setting their research agendas. Consequently, the Andrews LTER as well as the overall site program is a collection of individual-investigator projects tied together by a series of conceptual models. The committee found this to be an important factor in reviewing the data management and integration activities there.

At the heart of the data system for the Andrews site is the Forest Science Data Bank (FSDB). This database began to be developed during the International Biological Program, before the Andrews Forest was designated as an LTER site. It has benefited from the direct participation of many scientists interested in conducting regional research on the structure and function of the forest and stream ecosystems and their response to natural disturbances, land use, and climate change. Currently, 50 scientists from several institutions participate in this effort. FSDB houses 2,400 data sets from over 350 studies (databases) and adds data from about 20 new studies a year. The data are organized in 11 categories, such as hydrology and vegetation management. The total volume of the ground observation data is less than 300 megabytes, with approximately 200 gigabytes of remotely sensed data. Over $100,000, representing a significant fraction of the program's total research budget, is devoted to information management support for FSDB.

FSDB resides on a local-area network server. Local users have on-line access to the server and a set of coupled central catalogs. These catalogs contain information on the nature of the studies, their purpose and goals, their data collection activities and their periods, parameter lists, location information, experiment design, and many other relevant factors. The coupled catalogues allow search and cross-referencing for the purpose of locating the data sets that may be of potential interest to a researcher. Actual data and metadata (e.g., definition of a variable, minimum and maximum values) are stored in separate subdirectories for each study. New features built into the system allow automated export of data into the analysis systems, such as Geographic Information System or statistical analysis tools, for further processing.

The management of data in FSDB has certain characteristics that are fairly typical of ecological research. For example, data sets tend to be small and highly diverse, there is a tendency to keep data sets at the individual-investigator level, and the methodologies used in obtaining and managing the data are diverse and not necessarily standardized. As the LTER Program has progressed, the value of these disparate data sets has increased, not only for the originating principal investigators, but also for the co-investigators and other scientists, who have begun to integrate multiple data sets.

These factors have tended to help the development of FSDB. The managers of the data system stressed that the biggest incentive for scientists to use the system is improved access to one's own data, as well as better access to other researchers' data, both at the Andrews site and across all of the LTER Program's sites. Because of the diverse nature of the data sets in general and the long-term nature of the research, the data must be well documented to ensure not only that principal investigators can use the data, but also that future researchers can understand how the data were taken.

The existence of two complementary demands—the long-term nature of the data collection and the diverse nature of the data sets and collection methodologies—has led to the development of a sophisticated metadata management system. The committee was impressed by the time and effort spent by the Andrews LTER project in this area.

This improved access to the data by the researchers associated with the Andrews site, however, has also presented a problem of ownership to the data managers. While individual principal investigators have seen the advantages to obtaining their colleagues' data sets, they also have perceived the danger of unauthorized access to their data. Therefore, the managers of the database have built in a safeguard that allows a principal investigator to veto any use of his or her data. The amount of data that the data managers can actually release on their own authority is quite limited. Nevertheless, such restrictions have been replaced by federal regulations that require all data collected with federal money to be made publicly available no later than 2 years after collection. In reality, this issue has not proved to be much of a problem. Within the initial 2-year time frame, the only people who generally would know of and want the data set were researchers associated with the project. Therefore, the rules governing data distribution remained in effect without difficulties. During the few times when outside groups asked for data, the requests were accommodated.

The data collected in the early stages of the LTER Program, including the Andrews site, are more difficult to obtain along with adequate metadata. This situation has improved, not only because of the reasons given above, but also because an increasing reliance on mathematical models has encouraged the use and interpretation of a variety of data sets. The data management team as well as the principal investigators now try to anticipate data management issues at the beginning of each individual study project, including the incorporation of metadata support.

The committee supports use of the Andrews site approach that emphasizes the creation and electronic distribution of metadata catalogs as an appropriate first step toward better integration of the other LTER sites. This step in isolation, however, does not ensure evolution toward a fully accessible and optimal data system.

With regard to institutional issues, there appear to have been few problems between U.S. Forest Service and Oregon State University personnel in collaborating on this project. NSF has been very supportive and, in general, is gently pushing the other LTER sites to greater coordination, including intercalibration of data collection and study techniques. The local institutional aspects thus appear positive and headed in the right direction. The same features that encouraged good working relationships, as described earlier in this chapter, are responsible for helping to develop a good data management system as well.

With regard to NSF's overall management of LTER sites, interaction among sites is hoped for, but certain impediments get in the way of significant site interaction. First, each site needs to compete for funding every 6 years and, in a sense, is always in competition with other existing or potential sites. Second, until recently, there was no mechanism for funding to cooperate across sites. This, however, appears to be changing because of a realization that there is now a large body of data from different ecosystems in the United States and some means needs to be developed to provide access to these data sets.

NSF is now trying to encourage data sharing more actively among LTER sites. For example, the agency has funded an LTER data management center at the University of Washington to facilitate exchange among the principal investigators at the different sites. That consists primarily of a rapid means of e-mail communication, directories of addresses of other investigators, and the generation of a data index catalog that anyone can use to see what kinds of data sets are available and where they can be obtained.

NSF also is encouraging the development and interchange of data through funding allocations and equipment grants. For example, there is a small amount of money available to support intersite data management. In addition, NSF has facilitated the development of local-area networks, other wide-area networks, and high-capacity data storage and has provided funding for GIS equipment to help develop data management capabilities. No further technical standards beyond the Minimum Standard Installation for the LTER internetwork effort are anticipated for the near future.

LESSONS LEARNED

The Forest Science Data Bank is an excellent example of a scientific information management system that has been created and has gone through several evolutionary phases within an academic environment. During the early phases of development, the system evolution was dominated by the desires of individual researchers without major attempts at coordination, integration of functions, or identification and definition of high-level system requirements. The lack of an architecture that usually results from such an approach, along with the desire of scientists to minimize data entry and file storage costs, resulted in an unstructured system and led to difficulties in system maintenance and enhancement process. System upgrades proved to be especially costly and cumbersome, not because of the cost of the new hardware and software, or inconsistent cooperation between the researchers, but because local optimizations had led to structural flaws, such as absent or incomplete metadata, or lost and/or incomplete files. These and other similar deficiencies were difficult to detect, and when detected were difficult to correct.

Centralized information management support by a group of competent individuals grasping both science and data management issues has been the start of a new and successful phase in the evolution of FSDB. The activities of this phase have brought discipline to the collection and organization of the data and metadata and have improved users' access through relational catalogs, which can be searched and cross-referenced. The cost of these activities, however, is not trivial, running at about 20 percent of the total research budget.

Even though FSDB has made positive steps, it should not be considered a modern state-of-the-practice scientific data system. The system lacks a modern users' interface, has limited access capability, is made up of a large number of small data sets, and will probably continue to be costly to maintain and upgrade. Because of the small number of principal investigators (the primary users), these shortcomings have not posed a serious operational problem so far. The situation, however, could become an issue when more widespread access by other LTER sites is required. On the positive side, and as far as collection and organization of metadata are concerned, FSDB should be considered an excellent model. Considerable effort has been and continues to be devoted to the standard format and automated data entry procedures for metadata. These steps have led to less time-consuming efforts by the researchers and a better organized set of very useful metadata.

REFERENCES

Franklin, J.F., C.S. Bledsoe, and J.T. Callahan. 1990. Contributions of the Long-term Ecological Research Program. *Bioscience* 40(7): 509-523.

Harmon, M.E. 1992. *Long-term Experiments on Log Decomposition at the H.J. Andrews Experimental Forest.* USDA Forest Service General Tech. Rep. PNW-GTR 280, Portland, Ore.

Institute of Ecology. 1981. *Experimental Ecological Reserves: Final Report on a National Network.* The Institute of Ecology, Indianapolis, Ind.

McKee, A., C.M. Stonedahl, J. Franklin, and F. Swanson. 1987. *Research Publications of the H.J. Andrews Experimental Forest, Cascade Range, Oregon, 1948 to 1986.* USDA, Forest Service, Pacific Northwest Research Station, General Tech. Rep. PNW-201, Portland, Ore.

Michener, W.K., A.B. Miller, and R. Nottrott. 1990. *Long-term Ecological Research Network Core Data Set Catalog.* Belle W. Baruch Institute for Marine Biology and Coastal Research, University of South Carolina, Columbia, S.C.

5

The Carbon Dioxide Information Analysis Center

The Carbon Dioxide Information Analysis Center (CDIAC) at the Department of Energy's (DOE) Oak Ridge National Laboratory (ORNL) is internationally known and admired for its role in providing high-quality atmospheric data sets to the research community. These include time series measurements of carbon dioxide and methane at multiple stations around the world, as well as global estimates of the annual production of carbon dioxide from fossil fuel combustion and cement manufacture. In addition, CDIAC is active in efforts to "rescue" historical climate data that can provide useful comparisons with present data on trends in atmospheric conditions. One prominent example of this is a cooperative program with the Institute of Atmospheric Physics and the Institute of Geography in China.

While CDIAC does not directly engage in interfacing biological and geophysical data types, its data management program exemplifies the kinds of data gathering, quality control, documentation, and dissemination activities that are a necessary part of many data interfacing exercises. Depending on the nature of the interfacing effort, these activities could occur during the acquisition of geophysical and biological data or during the actual interfacing process itself. Despite past successes, the center's staff recognize that their data management model will not be adequate for meeting the challenges of processing and integrating larger volumes of data and doing so on shorter turnaround times. Because these challenges are common to the climate change research community as a whole, the committee believes that the following description of the center's data management approach will be widely applicable. There are explicit lessons to be learned both from its current success and from the challenges it faces in the future as it scales up its data management efforts.

CDIAC is a part of ORNL's Environmental Sciences Division. It was founded in 1982 by DOE to provide identification, collection, quality assurance, documentation, and distribution for information on the biogeochemistry of carbon dioxide and the effects of carbon dioxide on vegetation and on the Earth's climate. The scope of CDIAC was subsequently expanded to include related global change topics, such as other greenhouse gases and the effects of climate change on the environment.

Other programs not part of CDIAC, but within the Environmental Sciences Division, include the Atmospheric Radiation Measurement (ARM) archive, which will hold large data volumes (1 to 5 terabytes/year) related to general circulation models (specifically, the representation of clouds and of moisture, heat, and energy transfers therein) and derived from high-speed, real-time samplers. The Environmental Sciences Division also houses a NASA Distributed Active Archive Center, which focuses on ground-based field program data (e.g., carbon in soils, vegetation cover). In large part, this Distributed Active Archive Center was sited at ORNL because of CDIAC's past experience and success, which may be expected to be incorporated and extended into the ORNL DAAC's data management scheme.

ORNL is facing significant technical and organizational challenges as it attempts to implement the new functions associated with the ARM Program and the Distributed Active Archive Center. These challenges are representative of those faced by the global change research community as large volumes of data from new sources become increasingly available. ORNL's experience with CDIAC is relevant and valuable, but these two new programs are different in important ways. First, data volumes will be much larger than those with which CDIAC staff are accustomed to dealing. Second, these programs will focus on real-time rather than historical data. Third, ORNL will be serving a much larger audience and will not be as close to the user community as CDIAC's staff currently is. Finally, ORNL will not always have the luxury of time that is now available to CDIAC to build relationships, perform intense quality assurance and quality control, and produce value-added products.

VARIABLES MEASURED AND SOURCES OF DATA

CDIAC produces aggregate data sets that summarize global and regional production of greenhouse gases such as carbon dioxide and methane; trace gas measurements in the atmosphere and oceans; long-term climate records in addition to temperature (e.g., precipitation, clouds, atmospheric pressure, and storm climatologies); soil chemistry; coastal vulnerability to rising sea level; global distribution of ecosystem types; and the response of vegetation to elevated ambient carbon dioxide. These variables are derived from a variety of other direct and indirect measurements and estimates gathered from a range of sources. For example, the carbon dioxide emissions data sets are derived from United Nations energy production estimates, from Bureau of Mines cement data, and from DOE gas-flaring statistics (see ORNL, 1991).

DATA MANAGEMENT

This section describes CDIAC's strategy for the management of data. It shows that the center benefits from an unusual degree of freedom in its ability to select data sets for publication and to negotiate agreements with data sources. In addition, it reveals that the data management strategy depends for its success on the large amount of personal attention that each data set receives from the staff. These factors have been important reasons for CDIAC's success.

Selecting the Data Set

CDIAC staff stay abreast of current research issues by attending conferences, symposia, and workshops, by sponsoring workshops, and by interacting directly with researchers. Often, users will ask for information that does not yet exist. CDIAC's global carbon dioxide emissions data set is an example of a product that was created in anticipation of a need as well as in response to such a request. Another such product is the data set on coastal susceptibility to sea level rise. This data set uses Geographic Information System (GIS) technology to integrate data on sea level, erosion, coastline location, and elevation. It reflects scientists' increasing interest in using GIS as an integrative tool.

CDIAC prioritizes potential new data sets and then obtains feedback from sponsors and research groups. Political considerations sometimes influence the choice of what data to work on. The Chinese climate data project mentioned at the beginning of this chapter reflected a management decision to include a more globally diverse array of data. Sometimes a persistent principal investigator can influence the selection decision. Databases deemed to be of lesser scientific importance or whose credibility is in doubt because of methodology will get a lower ranking, whereas technically sound and scientifically important databases will be ranked higher. The size and source of the data are not important in the selection decision.

Contacting the Principal Investigator

CDIAC staff must convince the investigators to submit data to the center. This is because, with the exception of some DOE projects, the center does not have formal relationships that give it the "right" to acquire data. CDIAC staff emphasize that they will document the data, increase both the data's and the investigator's visibility, and remove the burden of responding to data requests. They also stress that the investigator will get full credit for the final data product, will have final sign-off authority on the data set, and can submit the data to the center in whatever format the investigator considers desirable. Because investigators can, and sometimes do, reject these offers, CDIAC staff must adopt a cooperative attitude. They may, for example, offer to wait for the data, while the investigator meets publication deadlines. The overall message is that any extra burden on the investigator will be minimized and that the center's involvement will result in a better product in the end.

Acquiring the Data

At the time that investigators submit data, CDIAC personnel attempt to get as much metadata as possible, including methods, reprints, contact names, and anecdotal information about the data. The contributing scientist is encouraged to send the data in whatever form is convenient. At this point, one person is assigned responsibility for the data set from start to finish. This expands the range of the staff's skills, because a variety of problems are common. In addition, a staff member will care more about a data set for which he or she has full responsibility than if oversight were fragmented. The lead staff person can draw on other expertise as needed.

Performing Quality Assurance and Quality Control

If data are submitted on hard copy, the CDIAC staff perform double data entry. For digital data, they perform virus checks and then make a backup. There is no standard quality assurance/quality control (QA/QC) methodology that is applied to all data sets, because each data set is unique, with its own peculiarities. Based on information from the investigator and past experience, CDIAC staff customize a QA/QC approach to each data set, depending on its characteristics. The operating assumption is that the submitted data are not clean. There are three elements that make this customized approach work: (1) ongoing interaction with the investigator to resolve problems; (2) the continuity that comes from having a single staffer with beginning-to-end responsibility for each data set;

and (3) experienced staff with scientific backgrounds relevant to the data sets.

The lead person for a data set develops a preliminary QA/QC plan and then presents this to other staff for discussion. Following this, the plan is reviewed by the investigator and other experts. The plan includes items such as key thresholds and relationships that must be internally consistent. The QA/QC plan usually is an effective starting point, but surprises sometimes occur that require subsequent improvisation. The plan often necessitates successive passes through the data, because some problems mask others that do not become visible until the problem in the "foreground" is corrected.

All corrections are discussed with the investigator before any changes are made to the data. If the investigator concurs, the change is made and noted in the documentation. If the investigator does not concur, the data are left as is, but the value is flagged as suspicious. After all the visible problems are corrected, the data set is sent out to be "beta tested" by researchers who perform analyses with the data in an attempt to uncover errors or discrepancies that slipped through the QA/QC process. This is a key part of the QA/QC process at CDIAC and provides an opportunity to evaluate critically the data from several different perspectives. The beta test step is based on the recognition that the in-house QA/QC process cannot realistically replicate all the data manipulations that an analyst would be likely to perform.

The data set of global emissions of carbon dioxide provides many examples of typical data quality problems. This data set required integrating four data sets that were not originally intended to be integrated—energy statistics, cement production estimates, gas flaring estimates, and population estimates. In this case, it was frequently necessary to create new data by analyzing or converting existing data. For example, gas flaring estimates for individual countries sometimes have to be estimated from crude oil production estimates and per capita emissions estimates. In addition, political considerations have obscured data or put constraints on how they could be used or reported. For example, some countries have been reluctant to publish raw population data, and the United Nations specified that CDIAC's data set on carbon dioxide emissions could not contain such numbers. However, the data set does contain total carbon dioxide production and the per capita production. The center's staff also were unable to resolve discrepancies in politically sensitive issues, especially for United Nations energy statistics. These data often do not agree with those from other sources, such as private industry or the Organization for Economic Cooperation and Development. However, because approximately 80 percent of carbon dioxide emissions come from about 20 percent of the countries, the CDIAC staff judged that such problems in relatively smaller sources are not critical. When the CDIAC staff first produced the data set, they reviewed United Nations data from 1950 and found many discrepancies and suspect values, because this was the first time these data had been critically reviewed. Problems included issues such as multiple entries per year per country, or a given country being shown as exporting more coal than it produced. As a result of this positive collaboration experience, the United Nations now utilizes CDIAC as a beta test site for its data sets before public release.

Documenting the Data

The center uses the "20-year rule"; that is, it prepares metadata that would make the data usable 20 years hence, when investigators who collected the data are no longer available for consultation. In the metadata, CDIAC especially emphasizes on limitations of the data and restrictions on possible uses. The documentation also discusses peculiarities and quirks that should be taken into account. In addition, it includes a hard copy of a subset of the data for validation purposes. This subset can be checked against the recipient's digital version to ensure that no problems have occurred during the transfer and loading of the data. Further, the documentation often includes one or more simple algorithms or derived variables to enable users to check on the integrity of the data set as a whole. For example, the documentation might contain the sum of a particular data parameter, added up over all records in the data set. Upon receipt of the data, the user could calculate this sum and compare it to the value in the documentation.

Both the data set and the documentation are reviewed by a team of independent reviewers. This review is made as rigorous as possible and is considered to be equivalent to a peer review of the data package and related metadata.

Distributing the Data

CDIAC staff distribute and publicize their data packages through as many avenues as possible. These include the NASA master directory; electronic bulletin boards; the CDIAC newsletter, *CDIAC Communications*, with more that 9,000 subscribers in 150 countries; university libraries; catalogs of CDIAC's data and information products; announcements sent to a network of newsletter and journal editors; and a mailing list compiled from conferences, sponsoring agencies, and past data requests.

The center sends out regular updates and special news items announcing new data products and revisions. The center also ensures that the investigators are kept up to date on requests for their data as well as on feedback about the data. Periodic surveys of the entire user community are performed, and these typically achieve a 50 percent response rate.

CDIAC keeps multiple copies of each data set on different media and at different locations. The National Technical Information Service is used as a method for preserving and disseminating the center's reports and data.

LESSONS LEARNED

A key feature of CDIAC's activities is the staff's understanding that successful data interfacing requires error correction and other quality control activities. Their experience shows clearly the importance of resolving discrepancies between data sets, clarifying ambiguities, investigating the implications of differences in measurement methods, backtracking from derived variables to the original raw data, and creating standardized measurements from a variety of sources. Unless these activities are performed thoroughly and accurately, data interfacing will not result in useful data sets.

As a consequence of working at this "hands-on" level with data, CDIAC staff identified several key prerequisites or premises that they felt were instrumental in their success in creating high-quality data sets. These premises reflect both technical and organizational factors and seem ideally suited to the scale of the center's data management activities to date. Many of these premises will be difficult to duplicate with the much larger volumes of data envisioned in the near future. Nevertheless, the center's staff have placed a high priority on developing ways to incorporate as many of these premises as possible in the expanded activities that will accompany their role as a data center. Each of the following paragraphs summarizes a distinct premise or prerequisite. Many of these are identical to those identified as essential to improving quality in manufacturing and service industries.

- *Strong commitment to service.* A strong commitment to service is a primary goal. The staff have identified their market as the research community and expend a great deal of effort to stay in touch with researchers to find out what they want. They have avoided intricate, high-technology systems and instead emphasize producing high-quality data and useful documentation. They focus on answering the question, "What kinds of data should be in this directory?" rather than on state-of-the-art methods of data transfer. They feel it is more useful to send out a high-quality data set on a tape than a substandard data set on a more advanced medium. While they are interested in responding to their more sophisticated users, they realize they also must remain accessible to many less well-trained users in developing countries. They do not see their role as being technology drivers.

CDIAC managers also place a high emphasis on fulfilling users' requests completely and in a timely manner. In becoming a World Data Center, they were concerned about the requirement that such data centers accept all data submitted to them for a certain area. A related concern was that the resulting large volumes of data could overwhelm their ability to continue emphasizing data quality and effective service. As a result they tripled their staff and upgraded their hardware in order to keep fulfilling their commitment to respond to users.

- *Collaborative mindset.* CDIAC personnel emphasize a collaborative mindset. They recognize that there is little reward for researchers to manage data for use by others and therefore try to relieve them of this burden. They will accept data in any format that is convenient to researchers and will work with them to make data submission easy. This close interaction with researchers also helps the center evaluate what kinds of data products would be useful or worthwhile to the research community. In contrast, other data center programs that mandate a single format for submitting data have experienced difficulties and have created a motivation for researchers to circumvent the program.

- *Full credit for data sources.* The CDIAC staff try to keep their methods and working relationships in accord with the research community's reward system. This effort creates additional incentives for researchers to provide their data and to participate in the sometimes complex and time-consuming QA/QC process. Data sources get full credit for the data, because they, not the CDIAC staff, are listed as authors on the data packages that the center produces. The staff negotiates with data sources in order to address concerns about others taking improper credit for the data. In some instances, they will agree to delay data submission until the source's analysis has reached a certain point or the results have been published. In addition, the staff recommend a citation format in the data packages, analogous to that for peer-reviewed journals, to help ensure that the sources get full credit.

- *No fee for services.* There is no charge for CDIAC's services, and this helps them build good working relationships with the user community. Not only does the center provide complete data sets with accompanying documentation, they also prepare regional subsets of data or customized combinations of specific data sets on request.

- *Emphasis on QA/QC and documentation.* CDIAC staff emphasize the value added that QA/QC and metadata represent. They argue that their data cleanup and documentation make data sets much more accessible and valuable to the user community. The operating assumption is that no data set is clean. Sufficient time and resources therefore are allocated for thorough beta testing of data packages. The center sends preliminary versions of data packages out to selected researchers, who then review the data from an analyst's perspective.

CDIAC staff take a long-term perspective, when necessary, in order to improve a source's data quality. For example, the center is involved in obtaining proxy climate records from China (e.g., early monsoons, rice harvest records). Staff are working with the aforementioned Institute of Atmospheric Physics and the Institute of Geography in China and have furnished them with PCs and data entry systems. The first data sets had numerous problems such as the minimum for a variable being greater than the maximum, 50 days of snow in a single month, precipitation of 0 while the qualitative data indicated a rainy month, and 0 used for missing. The project began in 1985, with the agreement signed in 1987. There was a long learning curve before the data quality improved. Data sets were ready to be published in November 1991. Although this is an extreme example, it is common for CDIAC to spend 1 to 2 years preparing a data set for publication.

- *Use of raw data.* CDIAC emphasizes the providing of original raw data, rather than derived or processed data. For example, the center's staff made a scientific case for obtaining the raw, instead of the derived, data when working on the data that were ultimately published as Numeric Data Package (NDP) 20, Global Grid Point Surface Air Temperature (Jones et al., 1991). There were four records for each time and place, but the temperature typically differed among the four records. Jones had used an algorithm to decide which was the "correct" record. Rather than simply providing these processed data, the center furnished the raw data to enable users to try different algorithms and compare their results with those produced by Jones. However, it took additional time to acquire and then work with the raw data.

- *Emphasis on proper staff training.* Based on the conviction that computer science skills alone do not provide the intuition needed for effective QA/QC, CDIAC's professional employees generally have a scientific background in addition to computer programming training. The center's location in Oak Ridge also has proved extremely valuable, because the staff have access to the expertise of a wide range of scientists when needed to help evaluate data.

- *Responsibility and rewards for staff.* CDIAC managers assign one staff person to have responsibility for each data set. This increases skill levels and makes sure someone has the "big picture." In addition, staff care more about "their" data set than they would if responsibility were fragmented. Staff are rewarded for data management skill and success and for soliciting feedback from the user community. These policies contribute to a low turnover of staff, which in turn retains learning in the organization and permits staff to continue improving their skills. This pool of experience also makes it easier to train new staff.

- *Added focus on nontechnical issues.* CDIAC staff recognize that data management and distribution are not just a technical exercise. Equal emphasis is given to organizational and motivational issues. A major factor contributing to the center's success is that it is operated as a long-term program with secure funding at a consistent level.
- *Ability to be selective in accepting data sets.* The center is not required to accept all data sets that may be submitted to them. Its ability to be selective means there is less danger of staff overload; as a result they can spend the time needed for intensive QA/QC and documentation.
- *Good working relationship with sponsor.* Finally, the center has a good relationship with its sponsoring agency, the Department of Energy. The staff identified a single person, Fred Kuminoff, as CDIAC's champion during its initial years. The current management maintains a commitment to providing data-related services and has helped focus CDIAC on a role it could fulfill effectively.

REFERENCES

Jones P.D., S.C.B. Raper, B.S.G. Cherry, C.M. Goodiss, T.M.L. Wigley, B. Santer, P.M. Kelly, R.S. Bradley, and H.F. Diaz. 1991. Numerical Data Package (NDP) 20, Global Grid Point Surface Air Temperature. Carbon Dioxide Information Analysis Center, Oak Ridge National Laboratory, Oak Ridge, Tenn.

Oak Ridge National Laboratory (ORNL). 1991. *Trends '91*. Thomas A. Boden et al., eds. Environmental Sciences Division, ORNL, Oak Ridge, Tenn.

Oak Ridge National Laboratory (ORNL). 1993. *Carbon Dioxide Information Analysis Center: FY 1992 Activities*. R.M. Cushman and F.W. Stoss, eds. Environmental Sciences Division, ORNL, Oak Ridge, Tenn.

6

The First ISLSCP Field Experiment

The overall goal of the International Satellite Land Surface Climatology Project (ISLSCP) is to improve our understanding of satellite measurements relating particularly to the fluxes of momentum, heat, water vapor, and carbon dioxide from land surfaces. The First ISLSCP Field Experiment (FIFE), which was a pilot study for further ISLSCP investigations, had the following specific goals (Sellers et al., 1988):

1. To determine whether our understanding of biological processes on the small scale, from microns to meters, can be integrated over kilometer scales to describe interactions appropriate for climate models; and

2. To determine whether biological processes (photosynthesis, evapotranspiration, etc.) or associated states (chlorophyll density, soil moisture, reflectance, etc.) can be quantified over appropriate scales for climate models.

FIFE was carried out over a 15 km × 15 km site in Kansas in 1987 and 1989. The site was the northwest portion of the Long-term Ecological Research (LTER) Konza Prairie Research Natural Area, a 3,400-ha tract of unploughed native tallgrass vegetation (LTER, 1991). The long-term monitoring program collected data through the entire 3-year period, complemented by four field campaigns in 1987 and a fifth campaign in 1989.

The lead agency for FIFE was NASA, with contributions from NSF, NOAA, the Department of the Interior, the Army Corps of Engineers, and the National Research Council of Canada. Ninety scientific proposals to participate were peer reviewed, and 29 groups were selected to undertake the field studies. The investigators then met twice yearly to agree on the experimental design and on follow-ups. The cost was about $100,000 per investigator per year. Operation of the FIFE data management system consumed about 10 percent of the budget. One of the indirect benefits of FIFE was the interactions that were generated between scientists working at the very small scale (ecologists, micrometeorologists) with those working at the pixel and larger scales.

A general description of FIFE was given by Sellers et al. (1988), while accounts of FIFE information systems have been published by Strebel et al. (1989, 1990a,b). The main scientific findings of FIFE were published in the *Journal of Geophysical Research* (Sellars et al., 1992), but individual research results are to be found in various other journals, for example, Hall et al. (1989, 1992), American Meteorological Society (1990), and Brutsaert and Sugita (1992).

VARIABLES MEASURED AND SOURCES OF DATA

The operational goals of FIFE were (Sellers et al., 1988):

1. The simultaneous acquisition of satellite, atmospheric, and surface data:
    (a) Satellite data (NOAA-9, NOAA-10, SPOT, Landsat, GOES);
    (b) Airborne radiometric, wind, and turbulence data (permitting, for example, the estimation of vertical heat fluxes from the calculation of eddy correlations);
    (c) Surface/near-surface fluxes of water vapor, sensible heat, momentum, and CO_2; and
    (d) Surface/near-surface states (e.g., leaf area index, soil moisture);
2. Multi-scale observations of biophysical parameters and processes controlling energy and mass exchange at the surface, the goal being to determine how these are manifested in "satellite resolution" radiometric data; and
3. Provision of integrated analyses through a highly responsive central data system.

The principal investigators had varying levels of experience to guide them in the operational design of FIFE, particularly with respect to the coordinated collection of data obtained from satellites, aircraft, and ground sensors (and by 29 different research groups). A major challenge that they faced was how data collected on various spatial scales could be combined to achieve the primary research and operational goals of FIFE. Furthermore, the actual relationships among most of the observed parameters and processes and their dependencies on scale were largely unknown (Strebel et al., 1992).

A typical example of the configuration of platforms with meteorological instruments used during FIFE included the NOAA-9 polar orbiting satellite, two aircraft, and a helicopter. At the same time, a network of ground-based stations was recording standard weather information, as well as fast-response heat flux, evapotranspiration, and momentum data (Sellers et al., 1988).

DATA MANAGEMENT AND INTERFACING

The FIFE data archive contains about 100 different types of data sets. These include satellite, aircraft, and ground-based (atmospheric boundary layer, vegetation, surface, and subsoil) observations. Totaling about 120 gigabytes, the volume of data collected by FIFE is much larger than usually encountered in typical single-investigator research. Also, while the volume of data is not large in comparison to some data archives (e.g., EROS Data Center), the diversity of data types presented some formidable management challenges. Also, data were received in both digital and analog form and in various formats, including hard copy floppy disks, and high-density magnetic tapes. During the initial planning phase of FIFE, considerable thought was given to the question of how best to organize the data archive, the net result being the FIFE Information System (Strebel et al., 1990b).

As described by Strebel et al. (1990a), the FIFE Information System "was an integrated 'organism' composed of people, hardware, and software," which, when working together, "were to facilitate the flow of data from multiple sources to multiple users." The system was driven by users' needs; that is, it had to "adaptively respond to a fluctuating mix of input data and output demands" (Strebel et al., 1990b). The initial users were the principal investigators, their graduate students, postdoctoral fellows, and technicians. There were a number of challenges to the smooth functioning of the system. Some users were highly computer literate, while others were "exposed during the experiment to a range, volume, and complexity of data and analysis capabilities an order of magnitude beyond their previous experience" (Strebel et al., 1992). Also, users' needs sometimes changed unpredictably during the course of the study. For instance, some principal investigators changed institutions, and others (or their graduate students) changed their focus in response to broadening perspectives, perhaps seeking to test hypotheses suggested in discussions with other scientists at users' meetings. Another challenge was that measurement designs were readily changed in the field, in contrast to the predetermined sensor designs and configurations typical of NASA satellite missions.

Recognizing at the outset that users' needs could change, the managers of the FIFE Information System emphasized flexibility; that is, the system was designed so that it could be used to answer a whole range of interdisciplinary questions, not just the largely single-discipline ones posed by individual principal investigators. Other desired attributes of the data management system were identified as follows (Strebel et al., 1990a):

- able to be constructed rapidly;
- capable of handling widely diverse data types;
- evolutionary;
- science directed;
- responsive to user needs; and
- capable of playing an active role in "quick looks" in the field and in quality assurance.

One of the operational goals of FIFE was to provide integrated analyses, and this goal was achieved. During the course of FIFE, the Information System was shown to be an effective integrator of the sometimes disparate data collection programs—satellite, aircraft, and ground-based. Some principal investigators were unaccustomed to meeting deadlines for delivery of quality-controlled data, but they came to recognize the value of rapid exchanges of data among all project scientists. In fact, the FIFE Information System was one of the most important reasons for the success of FIFE.

The entire FIFE data archive is being put on CD-ROMs (Landis et al., 1992). This data set collection includes almost all of the ground measurements and selected aircraft and satellite imagery. The FIFE CD-ROM series will make the data available to individual scientists on their desktop computers and is a cost-effective alternative to maintaining on-line archival storage. One disadvantage of a CD-ROM system (Landis et al., 1992), however, is that finding files is the slowest function of a CD-ROM drive. Although rapid advances in CD-ROM technology are eliminating this problem, speed and ease of access were gained on the FIFE prototype CD-ROM by storing the 6,000 point-data files in a highly structured directory tree that uses more than 2,000 directories.

The designers of the data management system were clearly aware of the importance of adequate documentation, or metadata. Quoting Landis et al. (1992),

> In most cases the documentation must stand on its own, with nothing else to explain the instruments used, the collection methods, problems encountered, or any peculiarities of the data themselves. These files

may be the only help available to some poor graduate student given the task of using a data set. Therefore the FIFE project has standardized its documentation format around a 15-section outline, with subsections for related items. Thus analogous information can be found under the same headings in all the documents.

The evidence to date is that the metadata system is working well, making it possible, for example, for scientists not involved in FIFE to use the CD-ROMs. For example, scientists interested in comparing heat and water vapor fluxes obtained from low-flying aircraft with values estimated from ground-based instrumentation are fully satisfied with the metadata packages.

In retrospect, one failure in the design of FIFE has been recognized, namely, the decision at an early stage not to include modeling studies as a component of the field experiments. Although monitoring and modeling are iterative processes, it was decided to fund modeling studies only after the data had been collected. Strebel et al. (1992) pointed out that the result in the case of FIFE was that some important parameters were not identified and observed; there was no method of undertaking integrated data quality assurance; there was only rudimentary ability to use "quick-look" data to identify important or interesting events happening in the field in order to redirect data collection efforts; and there was no framework to set overall data integration guidelines for data system design.

Other, less serious, problems included the following:

1. *Scale problems*. Investigators working at the microscale (leaf area index, point-based heat fluxes, and so on) found the spatial uncertainty in satellite data and in aircraft transects difficult to manage. In general, these sources of uncertainty need to be documented and accounted for.

2. *Notation problems*. Investigators in different disciplines sometimes used different conventions and names for similar kinds of measurements or procedures.

3. *Time problems*. The surface biology group used local time, whereas satellite measurements were timed by Greenwich Mean Time. In some cases, a time of day was not recorded, only the date.

4. *Researcher preferences in data analyses*. Despite prior agreement that data analyses should be undertaken in an integrated fashion, researchers often preferred to use their own analytical tools in preliminary, or quick-look, analyses and in quality control. In some cases, individual investigators were reluctant to spend time documenting their preliminary analyses, which could be important in later integrated assessments.

LESSONS LEARNED

Several major lessons were learned based on the views of the staff of the FIFE Information System:

1. *Geographic Information Systems.* Initially, it was hard to use the site reference scheme in FIFE, because of poor coordination among investigators. FIFE subsequently adopted a within-site location/documentation grid system, and all GIS activities were directed by the information system team. This procedure should be followed in future field studies.

2. *Correction and calibration.* There was varying rigor and documentation across the suite of instruments. Correction and calibration should be required for all instruments, following guidelines developed by the information system team.

3. *Quality assurance/quality control (QA/QC).* There was no formal quantitative data quality assessment for most FIFE data sets. To help avoid this problem, QA/QC plans should be required in proposals, and full documentation should be submitted with data sets. The project data manager should set QA standards or guidelines.

4. *Major processing tasks.* External contractors could not handle operational processing timelines or maintain consistent formats. All processing software, including format standards, should be developed or tested by the information system team before operational processing is attempted.

5. *Modeling/integrative science.* No modeling was undertaken at the front end of the experiment. Thus the FIFE Information System team had no framework for organizing data sets, setting priorities, or doing QA. The data system and the integrative modeling effort must be tightly coupled, with both being directed by one individual, perhaps the project data manager.

From a more general perspective, quoting Strebel et al. (1992), "If there is a single lesson that we have learned in building the FIFE Information System, it is that classical data base engineering technologies will never meet the needs of large-scale interdisciplinary scientific investigations."

Despite these reservations, the FIFE Information System must be considered a success. According to Strebel et al. (1992), the FIFE data set is "a baseline study that could be revisited, for many purposes, by global change researchers and others for decades." Indeed, new users are still coming forward. For example, FIFE has provided test data for the development of a satellite processing algorithm—well beyond the original scope of the experiment. There are several reasons for considering FIFE to be a success:

1. FIFE was able to involve the user community successfully from the outset.

2. The information scientists working within the FIFE data management system played a central role in the fieldwork and in the subsequent analyses and syntheses (e.g., with respect to scaling upward from point to pixel scales).

3. The subsequent CD-ROM and associated metadata system is user-friendly, making the data sets available to a wide variety of scientists for purposes quite different from those envisaged by the principal investigators.

Several follow-up field experiments (e.g., BOREAS) are planned for the 1990s, each with increased numbers of investigators and disciplines, increased funding, and increased volumes of data. As pointed out by Strebel et al. (1989), the FIFE experience points the way to handling a database of 500 to 1,000 gigabytes, the technology now being available "off the shelf." Strebel et al. (1989) conceive of future investigators "with high powered workstations supporting a suite of interactive database tools like those developed for FIFE. The principal investigators' workstations then become individual elements of the complete data system as distinct from a 'window' to a massive central facility."

Another important lesson to be learned from FIFE is that information scientists—that is, researchers who maintain scientific oversight for an information system—should play an essential role in the integration of data needed to carry out large interdisciplinary field experiments successfully. Finally, significant funding (10 to 20 percent of the total budget) should be dedicated to the data management activities.

REFERENCES

American Meteorological Society (AMS). 1990. *Proceedings Symposium on FIFE*. American Meteorological Society, Boston, Mass.

Brutsaert, W., and M. Sugita. 1992. Regional surface fluxes from satellite-derived surface temperatures (AVHRR) and radiosonde profiles. *Boundary-Layer Meteorol.* 58: 355-366.

Hall, F.G., P.J. Sellers, I. McPherson, R.D. Kelley, S. Verma, B. Markham, B. Blad, J. Wang, and D.E. Strebel. 1989. FIFE: Analysis and results—a review. *Adv. Space Res.* 9: 275-293.

Hall, F.G., K.F. Huemmrich, S.J. Goetz, P.J. Sellers, and J.E. Nickeson. 1992. Satellite remote sensing of surface energy balance: Success, failures, and unresolved issues in FIFE. *J. Geophys. Res.* 97: 19061-19089.

Landis, D.R., D.E. Strebel, J.A. Newcomer, and B.W. Meeson. 1992. Archiving the FIFE data on CD-ROM. Pp. 65-67 in *Proceedings IGARSS '92*. IEEE, New York.

Long-Term Ecological Research (LTER). 1991. *Long-term Ecological Research in the United States*. LTER Pub. No. 11. LTER Network Office, College of Forest Resources, University of Washington, Seattle. 178 pp.

Sellers, P.J., F.G. Hall, G. Asrar, D.E. Strebel, and R.E. Murphy. 1988. The first ISLSCP Field Experiment (FIFE). *Bull. Am. Meteorol. Soc.* 69: 22-27.

Sellers, P.J., F.G. Hall, G. Asrar, D.E. Strebel, and R.E. Murphy. 1992. An overview of the First International Satellite Land Surface Climatology Project (ISLSCP) Field Experiment (FIFE). *J. Geophys. Res.* 97: 18355-18371.

Strebel, D.E., J.A. Newcomer, J.P. Ormsby, F.G. Hall, and P.J. Sellers. 1989. Data management in the FIFE information system. Pp. 42-45 in *Proceedings IGARSS '89*. IEEE, New York.

Strebel, D.E., S.G. Ungar, J.P. Ormsby, J.A. Newcomer, D. Landis, S.J. Goetz, K.F. Huemmrich, D. van Elburg-Obler, P.J. Sellers, and F.G. Hall. 1990a. The FIFE information system: Support of interdisciplinary science. Pp. 140-147 in *Proceedings Symposium on FIFE*. American Meteorological Society, Boston, Mass.

Strebel, D.E., J.A. Newcomer, J.P. Ormsby, and P.G. Sellers. 1990b. The FIFE information system. *IEEE Trans. Geosci. Remote Sensing* 28: 703-710.

Strebel, D.E., J.A. Newcomer, and K.F. Huemmrich. 1992. Data integration: The FIFE information system perspective. Unpub. ms. prepared for the USNC/CODATA Committee for a Pilot Study on Database Interfaces. 25 pp.

7

The California Cooperative Oceanic Fisheries Investigation

The California Cooperative Oceanic Fisheries Investigation (CalCOFI) program is an example of a long-term, broad-scale research and monitoring program that has been intentionally interdisciplinary from its outset. A major goal of the program has been to describe and understand the relationships between biological patterns and physical oceanographic/climate processes. This is consistent with the other case studies, which focus on efforts to integrate data from several different sources.

There is a long history in California of monitoring fisheries and fish catches, beginning in 1914 with the establishment of the California Department of Commercial Fisheries. The mission of this agency was to collect fisheries statistics, develop improved catch and processing methods, and study life histories of commercially important stocks (Hewitt, 1988). Throughout the 1920s and 1930s the department performed research on stock sizes and distribution, as well as year class abundance. After World War II, state fishery agencies in California, Oregon, and Washington, along with similar agencies in British Columbia, formed the Pacific Marine Fisheries Commission. The original goal of the commission was to study the sardine fishery. However, when that fishery collapsed in 1947, the commission turned to other fishery stocks.

The collapse of the sardine fishery was the precipitating event for the ultimate establishment of the CalCOFI program. After this collapse, the California legislature established the Marine Research Committee, which included representatives of the commercial fishing industry and the California Department of Fish and Game. In 1948 the Marine Research Committee established the California Cooperative Sardine Research Program, with the goal of studying the distribution and natural history of sardines, their availability to the commercial fishery, fishing methods, and the physical, chemical, and biological oceanographic processes influencing sardine populations in the coastal waters of California. Members of the program included the California Department of Fish and Game, the Federal Bureau of Commercial Fisheries, Hopkins Marine Station, the California Academy of Sciences, and Scripps Institution of Oceanography. The program was renamed the California Cooperative Oceanic Fisheries Investigation in 1953, when its scope was expanded to include other species besides the sardine.

Over time, the program's objectives evolved until, by 1960, they were primarily to understand the factors controlling the abundance, distribution, and variations of pelagic marine fishes. A major emphasis was on comprehending the physical and biological oceanographic processes affecting marine life in the California ocean current system as a whole (Baxter, 1982).

Even though the 1976 Fisheries Conservation and Management Act (FCMA) gave the federal government management authority over commercial fisheries in the exclusive economic zone (3 to 200 nautical miles from shore), the National Marine Fisheries Service (NMFS), the California Department of Fish and Game, and Scripps decided in 1979 to continue the CalCOFI program as a long-term marine resources monitoring and research program (Radovich, 1982).

VARIABLES MEASURED AND SOURCES OF DATA

The CalCOFI program measures a variety of biological and physical oceanographic variables. Plankton and neuston tows are used to collect ichthyoplankton, invertebrate zooplankton, and phytoplankton. Rapid postcruise measurements of zooplankton biomass are made, and some investigators work up the samples to greater levels of taxonomic detail. Primary production is measured daily; continuous measurements of temperature, light, chlorophyll, and dissolved oxygen are taken; and Acoustic Doppler Current Profiles of current measurements and acoustic backscatter are collected. Chlorophyll, phaeophytin, salinity, dissolved oxygen concentration, nitrate, nitrite, phosphate, silicate, and water transparency data (to depths exceeding 500 m) are taken from a 20-place salinity-temperature-depth (also known as CTD) rosette on a grid of about 70 stations quarterly during the year. These measurements have been taken regularly for the past 44 years, using the technologies current at the time. Each spring, special egg and larval surveys are undertaken to determine the spawning biomass of certain commercially significant species of fishes.

In addition to these simultaneous and synoptic measurements, several cooperative measurement programs are commonly carried out on CalCOFI cruises by government agencies and by state and university scientists. In summary, the CalCOFI program was designed as a self-contained entity, with its chief focus on investigating relationships among the biological and physical processes under study. Thus, the program depends primarily on its own data (see CalCOFI, 1992).

DATA MANAGEMENT AND INTERFACING

As described below, the CalCOFI program was interdisciplinary in nature from its inception. The interfacing of biological and physical oceanographic data was thus integral to its success. This emphasis was reflected in the design of the sampling and measurement program, the staffing of the field and laboratory teams, the scheduling of periodic workshops to share information across disciplinary boundaries, and the way responsibilities were divided among the participating agencies. As a result, the CalCOFI program avoided many of the data interfacing problems encountered in the other case studies.

Science

Because the collapse of the sardine fishery was a broad-scale event that apparently had not been previously observed, the designers of the CalCOFI program were forced to consider systemwide mechanisms in their search for explanations of this event. From the beginning, the program's central question was, What are the broad-scale, long-term processes that drive temporal variability in fish populations? As a result of this focus, the CalCOFI program was interdisciplinary in nature, stressing the relationship between biology and physical oceanographic processes. At the time, during the late 1940s and early 1950s, the inclusion of physical oceanography in a fisheries investigation was a unique idea, and one that is credited to the influence of Harold Sverdrup. Sverdrup was also instrumental in designing the interdisciplinary graduate curriculum at Scripps Institution of Oceanography, where students in all earth science disciplines take a broad range of courses. The CalCOFI program thus had available an ongoing source of students and researchers with an ingrained interdisciplinary perspective.

In addition to spurring an interdisciplinary mindset, the central research question described above drove other important design decisions. There was an initial, fundamental, and long-lasting agreement among the participants to base the program on a long time series of data. This required frequent sampling in order to distinguish the relative importance of events occurring on different time scales. As described below, sampling therefore occurred monthly for the first 14 years of the program. In addition, because successful long-term analysis depends on using the same measurements over time, this agreement led to an emphasis on using relatively simple, but accurate and verifiable, measurements that would remain valid over the long term. The program consistently avoided sophisticated, state-of-the-art approaches where methods were likely to change rapidly and create intercalibration problems.

Because of the nature of the sardine collapse, the program was forced to direct its attention to broad-scale processes and sampled what at the time was an extremely large area. Between 1949 and 1963, monthly cruises sampled nearly 300 stations from the California/Oregon border to Cabo San Lucas at the tip of Baja California, and out to 400 nautical miles from shore. After 1963, these cruises were carried out quarterly rather than monthly. From 1966 to 1984, sampling occurred only every third year, but since 1984 sampling has occurred every quarter, from Point Concepcion to San Diego. By 1984 the program's scientists had shown that large-scale, low-frequency events were most important and that the smaller-scale sampling program was adequate to detect these events and their effects on the system. A critical part of the program's design was that biological and physical sampling occurred on the same temporal and spatial scales, thereby making it easier to look for relationships between the two kinds of processes.

Ensuring data quality has been a consistently high priority for the CalCOFI program. According to the briefings received by the committee, exacting standards were established early on by Hans Kline and were institutionalized and updated over the years. Problematic data points have been systematically examined and documented on "Form 8: Investigation of Doubtful Data," which is kept in the data file for each sample. Much of the success of the quality control effort can be attributed to the program's organizational features, which are described in the next section.

Organization

Just as the CalCOFI program from its outset was scientifically interdisciplinary in nature, it was also administratively cooperative. When the California legislature passed legislation in the late 1940s to set up the program, it did not designate a lead agency or appoint an overall director. Decisions were to be made by the participants acting cooperatively as equals. At present, there is still no formal locus of control; the program is run by a three-person committee made up of the director of the Marine Life Research Group at Scripps, the director of the NMFS Southwest Fisheries Center in La Jolla, and a representative of the California Department of Fish and Game. There is a single coordinator who works for this committee and whose job is to keep all participants effectively communicating.

The basic agreement among the three agencies that manage the program is renewed every 5 years. Administrative decisions are made by the committee on a cooperative basis. Scientific decisions are made by the principal investigators themselves, who often must balance competing priorities against severe budgetary constraints.

The fact that this cooperative decision making occurs successfully is due to several other features of the program's organizational structure and functioning. For example, an important feature of the program's enabling legislation was that the three main institutions—Fish and Game, Bureau of Commercial Fisheries (later National Marine Fisheries Service), and Scripps (later as part of the University of California)—were deliberately kept separate. Thus, operational budgets, staff, facilities, and other infrastructure were not joined, but remained under the control of each agency. This, and the fact that each agency received a fair share of program resources, minimized the likelihood of turf battles among the participants.

In addition, each agency had a primary mission that was distinct from that of the others. NMFS focused on studying fish eggs and larvae, Fish and Game on adult populations and catches, and Scripps on relationships between the physical and the biological environments. This arrangement contributed to scientific pride of ownership and lessened competition among the agencies. However, all participants shared a common purpose reflected in the program's central research question, which could be answered only by combining data from all three agencies. This structurally reinforced cooperation was facilitated by frequent informal conferences where investigators shared data, opinions, and arguments. Over the long term, these conferences contributed to the formation of fruitful working relationships among scientists from different agencies and different disciplines. In addition, the data collected through CalCOFI-sponsored research have always been made freely available to the broader research community.

While each agency maintained control over its own personnel and infrastructure, the crews for the CalCOFI cruises were provided by all the agencies. This, along with other mechanisms designed to foster communication, led to a greater degree of mutual respect and scientific interaction. At present, each agency commits some permanent staff to the program; as a result, the data collection and interpretation are highly integrated. The same staff who gather data on the cruises also process samples and data in the laboratory. This provides an automatic feedback mechanism in the data quality control process. The involvement of the same people in the entire data path from sample collection to report preparation is singled out by program scientists as a key reason for the high quality of the program's data.

Results

The CalCOFI program has produced one of the most consistent and highest-quality long-term data sets available for investigating the relationships between biological and physical processes. These data have been instrumental in three kinds of insights. First, they helped show that most of the temporal and spatial variability in the California ocean current system is contained in the low-frequency end of the spectrum. That is, the most important changes occur infrequently and over large areas. Second, they helped identify the linkages between biological changes and broad-scale shifts in water masses. Third, they helped reveal the connection between the shifts in coastal water masses and the periodic global-scale El Niño/Southern Oscillation phenomenon.

LESSONS LEARNED

The CalCOFI program has produced important results with a program design developed in the late 1940s around relatively simple parameters. This achievement is a tribute to the insight of the program's original scientists and an indication of the value of long time series of coordinated biological and physical measurements. However, the program's success is equally attributable to its organizational features. In combination, the program's scientific and organizational traits provide useful lessons for other attempts to interface disparate data types in order to examine complex processes. These lessons include:

- Build the program around simple, yet challenging questions that cut across discipline boundaries and foster a shared purpose among program scientists.
- Create an explicit interdisciplinary focus by framing research questions so that each discipline requires data from other disciplines to achieve its goals.
- Keep the program's guiding principles simple, both scientifically and organizationally, in order to maximize the flexibility and adaptability needed to pursue interdisciplinary problems.
- Involve scientists from all disciplines in shared decision making in order to minimize discipline-related turf battles.
- Keep the participating institutions' formal responsibilities clear

and uncluttered and keep bureaucracy to an absolute minimum. These precautions will assist in data interfacing, because resolving interfacing problems often requires cutting across organizational or bureaucratic boundaries.

- Coordinate sampling and measurement designs across disciplines to reflect a few simple and clear criteria that will support interfacing.
- Create structures, such as workshops, conferences, and shared field programs, that will promote cross-discipline working relationships.
- Emphasize the importance of data quality as a key prerequisite for successful data interfacing and productive interdisciplinary research.

REFERENCES

Baxter, J.L. 1982. *The Role of Marine Research Committee and CalCOFI.* CalCOFI Rep. 23. Scripps Institution of Oceanography, University of California, San Diego.

California Cooperative Oceanic Fisheries Investigation (CalCOFI). 1992. *Data Report: Physical, Chemical and Biological Data.* Scripps Institution of Oceanography, University of California, San Diego.

Hewitt, R. 1988. *Historical Review of the Oceanographic Approach to Fishery Research.* CalCOFI Rep. 29. Scripps Institution of Oceanography, University of California, San Diego.

National Research Council (NRC). 1990. *Monitoring Southern California's Coastal Waters.* Committee on a Systems Assessment of Marine Environmental Monitoring. National Academy Press, Washington, D.C.

Radovich, J. 1982. *The Collapse of the California Sardine Fishery. What Have We Learned?* CalCOFI Rep. 23. Scripps Institution of Oceanography, University of California, San Diego.

8

Interfacing Diverse Environmental Data—Issues and Recommendations

As described in Chapter 1, addressing many of the questions central to environmental research and assessment, and global change research in particular, requires combining geophysical and ecological data. Although this can be difficult, the success and cost-effectiveness of these research and assessment efforts depend significantly on the degree to which data interfacing issues are explicitly confronted. This chapter presents a working definition of data interfacing and describes in detail the technical and organizational barriers that impede it, including the barriers deriving from characteristics of data, from users' needs, from organizational interactions, and from information systems considerations. Specific recommendations also are provided. The chapter ends with a list of 10 Keys to Success, which are based on the committee's review of the case studies. These fundamental, generalized guiding principles should help practitioners to systematically respond to the challenges identified.

Real-world illustrations of problems and solutions relevant to data interfacing are used as examples throughout this chapter. Some of these are drawn from circumstances or applications that do not directly involve interfacing geophysical and ecological data. Examples of this sort were chosen because they effectively exemplify important elements or principles that are pertinent to such interfacing. Indeed, many of the challenges posed by interfacing these two data types are common to many other situations.

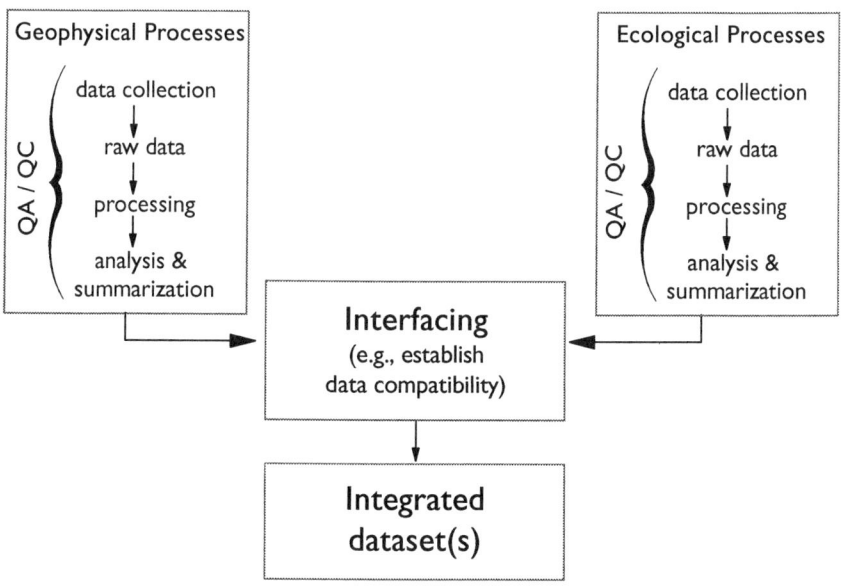

FIGURE 8.1 Generalized representation of the processes involved in interfacing geophysical and ecological data.

THE PROBLEM AND ITS CONTEXT

As defined in Chapter 1, interfacing of geophysical and ecological data is the coordination, combination, or integration of such data for the purposes of modeling, correlation, pattern analysis, hypothesis testing, and field investigation at various scales (Figure 8.1). The data being interfaced can be products of a single, integrated study or can be derived from several studies performed at different times or places. Similarly, the data could have been collected with the interfacing effort in mind, or for other purposes entirely. This deliberately broad definition of interfacing is intended to fit as many situations as possible. As discussed in greater detail below, the specific questions scientists will ask and the ways in which they will therefore endeavor to integrate data are often ill-defined and constantly changing (see Box 8.1). As a result, no single narrowly framed definition and no mechanistic prescription or solution will be of much lasting use to scientists contending with the problems related to interfacing.

At its simplest level, interfacing involves the identification, accessing, and combination of data. However, in practice these seemingly uncomplicated activities can be technically complex, stretching the limits of existing knowledge and the capabilities of available hardware and software

> **BOX 8.1**
> **Complex Questions Require Data Interfacing**
>
> Many fundamental questions about how ecosystems respond to forcing by larger-scale variables can be answered only by analytic techniques that use geophysical and ecological data together. The following examples from our case studies provide some sense of the range of such questions currently being addressed.
>
> - The California Cooperative Oceanic Fisheries Investigation (CalCOFI) program used a variety of correlation, pattern analysis, and time series methods to look for relationships between mesoscale shifts in oceanic current systems and biological communities. It was successful in showing how these regional oceanographic and biological changes were linked to the larger-scale El Niño/Southern Oscillation phenomenon.
> - The First ISLSCP Field Experiment (FIFE) study focused on the mass and energy balances at the land surface/atmosphere boundary and on the physical and biological processes that control them. The study combined ground-based, helicopter, aircraft, and satellite observations at several scales in order to develop and validate models that would allow surface climatology to be predicted from satellite measurements.
> - The National Acid Precipitation Assessment Program's (NAPAP) Aquatic Processes and Effects studies collected a large variety of environmental data, including precipitation rates, rainfall chemistry, rates of surface water acidification, and potential effects on aquatic biota. The end products sought were models to predict future scenarios of surface water acidification.

systems. In addition, the very act of interfacing frequently requires crossing disciplinary and administrative boundaries, thereby adding another level of complexity to the process. Interfacing therefore can best be understood as occurring in a series of overlapping contexts, both technical and organizational. Effective solutions must address and accommodate all of these relevant contexts.

Interfacing efforts can be confounded by a variety of obstacles (see Mathews, 1983; Henderson-Sellers, 1990). The ones described in this chapter are typical of most situations involving data management and data analysis on complex data sets. However, the challenges facing global change research and other large, interdisciplinary environmental research programs are extreme because of the massive volumes of data, the broad (up to global) geographic scale, the temporal scale, the variety of natural and anthropogenic processes included, the scope of modeling efforts, the numbers of organizations involved, and the evolving nature of the research itself. Consequently, the repercussions of not addressing the barriers described below are correspondingly more significant and more severe than for more traditional single-discipline applications.

ADDRESSING BARRIERS DERIVING FROM THE DATA

Data interfacing efforts must sometimes confront the misperception that once data are in digital form and in a common format, interfacing is simply a matter of merging two or more data sets. As Townshend and Rasool (1993) have pointed out, "collecting data globally does not of itself create global data sets." There is a series of technical pitfalls and obstacles that must be considered and resolved for data interfacing, even on regional scales, to produce scientifically meaningful output. Some of these stem from relatively simple discrepancies among data types and can be dealt with in a straightforward manner. Others, in contrast, reflect fundamental theoretical or "cultural" differences in the ways that ecological and geophysical studies are conceived and carried out. Barriers that arise from these more fundamental differences involve, among other things, the size and complexity of ecological versus geophysical studies, their spatial and temporal scales, the numbers and kinds of variables measured, the role of models in study design and analysis, and traditions of funding and project administration.

Spatial and Temporal Scale

Some of the most apparent barriers revolve around issues of scale. Geophysical studies are more likely to cover continental- and global-scale areas sampled at lower spatial resolution (Rasool and Ojima, 1989; Sellers et al., 1992a). In contrast, ecological studies generally tend to involve ground-based and closely spaced sampling of smaller areas over relatively short time periods. For example, a recent review of about 100 field experiments in community ecology revealed that nearly half were conducted on plots no larger than 1 m in diameter (Kareiva and Anderson, 1988). There is, in fact, only one widely used global data set in the ecological realm, the Global Vegetation Index produced by NOAA at a spatial resolution of 15 to 20 km (Townshend and Rasool, 1993). This stems, in part, from a tradition in ecology of studies on single species and from ecology's roots in natural history studies performed by individual investigators (Worster, 1977). It also reflects an emphasis in the conduct of ecological studies on labor-intensive field and laboratory techniques that preclude sampling over broader spatial scales. For example, in the FIFE study, data on canopy-leaf-area index, green-leaf weight, dead-leaf weight, and litter weight had to be collected by hand from relatively small study plots. This could not be avoided, even though the study was designed from the outset to integrate ecological and geophysical data over larger areas.

Because of such differences in study design, ecological data often must be smoothed or averaged in order to match the coarser spatial and temporal scales characteristic of geophysical data and models. This occurs, for example, when Geographic Information Systems (GIS) are used to merge remotely sensed areal data (usually geophysical) with attribute data (usually ecological) from specific points on the ground (Elston and Buckland, 1993). This kind of averaging is a key step in the latest generation of integrated global climate models (e.g., Wessman, 1992; Baskin, 1993). However, such averaged data may not truly be representative of heterogeneous ecological communities. This is an important shortcoming when heterogeneity is a vital component determining an ecosystem's response to a changed environment. In fact, Holling (1992) points out that spatial heterogeneity, or lumpiness, is of primary interest to ecologists.

Such differences in spatial scale also are related to the kinds of processes each field considers important, and the range of spatial and temporal scales across which they can be integrated. Most ecological studies thus attempt to focus on well-bounded community types and the actions of individual species or groups of species within these. Even studies that, by ecologists' standards, cover large areas (see Box 8.2) are fairly restricted compared to the global scope of many geophysical investigations. In addition, Wiens (1989) suggests that ecologists have been slower than atmospheric and earth scientists to address issues of scale. These other sciences (e.g., Clark, 1985) have a longer history of linking physical processes from local to global scales. Further, most ecological models function at a single scale (Ustin et al., 1991) or do not explicitly address scale (Wiens, 1989).

There are, of course, exceptions to this generalization about the spatial scale of ecological studies. For example, there is a long-standing tradition in ecology of interest in global patterns of community diversity and the body sizes of individual organisms, and more recent concerns about biodiversity (e.g., May, 1991; Jackson, 1994 a,b) and sustainability of the biosphere (Lubchenco et al., 1991) encompass a global perspective. However, none of these concerns has to date required the interfacing of large amounts of data from different sources.

The differences in spatial scale between geophysical and ecological studies are paralleled by analogous issues of temporal scale. Long-term time series of ecological data are relatively rare. This may make it difficult or impossible to create integrated data sets that focus on long-term changes in coupled ecological-geophysical systems. Where historical ecological data are available, they are more likely to represent data from several studies carried out independently over the period of research interest. Long-term ecological data sets of broad spatial extent are thus more likely to result from the combination of data from several sources. This in turn requires solving quality control, metadata, and data integra-

> **BOX 8.2**
> **Ecological and Geophysical Scales Differ**
>
> While the majority of ecological studies focus on relatively restricted spatial and temporal scales, other have a larger perspective. The committee examined two of these, the NAPAP and CalCOFI programs, as case studies (see Chapters 3 and 7). Other areas of study that attempt to link ecological processes across local to regional scales are described below. Each illustrates ways in which ecological and geophysical data might be interfaced. Even though large by ecologists' standards, they are small in comparison with may geophysical studies.
>
> Ecologists have used a wide variety of paleoecological data to explore the ways in which vegetation communities responded to past climate change, especially during and after the most recent glaciation. Most of these studies have concentrated on regions within Europe and North America (e.g., Davis, 1981; Cole, 1985; Pennington, 1986; Webb, 1987; Foster et al., 1990). A central concern in these studies is to understand how the ecological requirements and characteristics of individual vegetation species contribute to regional patterns of community change over time.
>
> Marine ecologists have expanded their understanding of how intertidal communities are structured by including oceanographic processes in their studies. Connell (1985), Gaines and Roughgarden (1985), and Roughgarden et al. (1985, 1986) showed that the importance of predation and competition for space within the intertidal community depends on the numbers of larvae available for settlement. This in turn depends on regional processes, such as currents and upwelling, that extend far beyond the intertidal zone.
>
> Forest ecologists are attempting to use simulation models that incorporate the birth, growth, and dynamics of individual plants to understand how vegetation would change on regional and global scales in response to climate change (Shugart et al., 1992). Such models include the physiological responses of individual plants to specific environmental variables. They also include somewhat broader-scale changes in community composition in response to physical disturbance as well as to environmental change.
>
> Long-term studies in the Chesapeake Bay have examined how regional land use, hydrology, waste discharge, and natural ecological processes interact to affect important estuarine resources. These studies were based on a systems approach that depended on interfacing data on many different aspects of the estuary (NRC, 1988).

tion issues stemming from methodological differences between the data sets.

In addition to these differences in study design, the behavior of ecological systems can confound the interfacing of geophysical and ecological data. For example, environmentally induced changes in ecological systems often occur with time lags of varying lengths (e.g., Cole, 1985; Lewin, 1985; Pennington, 1986; Davis, 1989; Loehle et al., 1990; Steele, 1991). This can make it difficult to determine which ecological and geophysical data should most appropriately be interfaced. Such time lags may require interfacing data from the same location or region, but from different times. In many cases, such data may be difficult or impossible to find. Research into the long-term response of vegetation to climate change (e.g., Davis, 1981, 1989; Cole, 1985; Pennington, 1986; Webb, 1987; Foster et al., 1990) has also shown that species respond individualistically, that is, that communities do not respond uniformly as entire units. This means that interfacing studies may provide misleading or inconsistent results when using "representative" species as indicators of ecological response to geophysical variables.

Reflecting such differences in temporal scale, the time steps in the models that represent geophysical and ecological systems can be quite distinct. For example, general circulation models recompute winds and temperature every 20 minutes for each grid cell. In contrast, ecological models of vegetation change use monthly to yearly, or even decadal, time steps (Baskin, 1993), depending on the kind of response being modeled.

These scale-related problems stem from the fact that fundamentally different kinds of processes are at work in geophysical and ecological systems. They also arise from the fact that ecological systems can be viewed at many different scales and from many different perspectives. None of these is the only "correct" one (O'Neill, 1988; Wiens, 1989; Levin, 1992), and each is based on different choices about which underlying processes and mechanisms to look at. This complexity, of course, affects choices about what kinds of data should be selected for interfacing. More complex problems that cut across several scales are complicated by the fact that variables and processes may or may not change in concert across scales (Wessman, 1992). Certain kinds of measurements in ecological systems may be correlated at one scale, but appear unrelated or negatively correlated at another (Wiens, 1989). In addition, sampling at intervals that are too widely spaced in space and time often fails to capture important aspects of a system's underlying variability. In such instances, the well-known problem of aliasing can lead even sophisticated analysis approaches to falsely identify trends. Thus the scales at which different kinds of data are collected can constrain or even predetermine the relationships among ecological and geophysical variables. As a result, data collected at one scale cannot necessarily be used to represent processes at another scale (through averaging or subsetting). This means that scientists engaged in data interfacing efforts must exercise extreme care when attempting to integrate data across different scales. Merely merging ecological and geophysical data by rote without seriously considering scale-related issues and their implications could result in spurious relationships and misleading analysis results. Unfortunately, there are no well-developed guidelines to assist in such efforts, although hierarchy theory (O'Neill, 1988) is a promising conceptual approach for identifying ecological scales that maximize predictive power.

Developing an integrated conceptual model of the systems being studied will assist in defining methodological and data management problems particularly with respect to spatial and temporal scales. For example, in the committee's NAPAP case study the decision to measure acid neutralizing capacity (ANC) was based on a model of how lake chemistry works. ANC turned out to be a key parameter in the survey of lake sensitivity.

The committee recommends that careful thought be given, in the planning for interdisciplinary research, to the implications of different inherent spatial and temporal scales and the processes they represent. These should be discussed explicitly in project planning documents. The methods used to accommodate or match inherent scales in different data types in any attempts to facilitate modeling and analysis should be carefully evaluated for their potential to produce artificial patterns and correlations.

Preliminary Data Processing and Statistical Uncertainty

A wide range of models and data processing algorithms typically are used in the development of ecological and geophysical data sets. Such preliminary processing is used for quality control and data cleanup, for data summarization and classification, and for extracting higher-level information from raw data (see Box 8.3). Raw data therefore are rarely used

---

BOX 8.3
Preliminary Processing Affects Data Compatibility

A wide range of preliminary processing methods are used to convert raw ecological and geophysical data to a usable form. Many of these can significantly affect later efforts to interface disparate data types.

- Raw images from satellite or airborne sensors must be processed to remove distortion and degradation that arise from a variety of sources (Geman, 1990; NRC, 1991, 1992; Geman and Gidas, 1991; Simpson, 1992). Estimating the true image intensity at each point often involves sophisticated spatial statistics that use adjacent data points to estimate the degree of distortion in the raw signal.
- In many studies, raw data are used to generate maps showing the distribution of variables such as age classes of trees in temperate forests, soil types, or ranges of sea-surface temperature. The resultant data sets no longer contain the original raw data, which can make it difficult if not impossible later to combine data sets with different class limits.
- In the Sahel case study, vegetation density was represented by a vegetation index. After several attempts using different alogorithms, a suitable index was derived by taking the difference between the reflectance of visible and near-infared bands and dividing this by the sum of the reflectances of the two bands.

directly, with the result that assumptions (both implicit and explicit), biases, and various kinds of statistical error are unavoidably built into each data set.

These built-in features of the data have two important kinds of implications for data interfacing. First, they mean that data interfacing involves more than just combining the tangible aspects of the data, such as formats and data values. It also necessarily involves identifying, understanding, and accommodating the assumptions, perspectives, value judgments, and decisions inherent in each data set. In simpler terms, all data sets cannot be all things to all people. For example, Townshend and Rasool (1993) list the various data products currently being derived from NOAA's Advanced Very-High Resolution Radiometer (AVHRR) data. Each product responds to the specific needs of a different subset of users, and these products are not readily interchangeable, even though all are derived from the same basic raw data.

Second, all preliminary processing and derivation steps are associated with some kind and amount of statistical error or uncertainty. Data interfacing, whether it involves combining separate estimates of the same variable, different variables, summary statistics, derived spatial data, or data that incorporate subjective judgments, represents another source of statistical error (NRC, 1992). For example, investigators focusing on point-based data in the FIFE program were unprepared to deal with the registration accuracy problem when requesting "their" pixel of AVHRR data. Plus or minus one pixel may mean a 5-km uncertainty in location, whereas the ground-based instruments were sensitive at 100-m scales. A more complicated problem arose when FIFE investigators attempted to associate averaged flux measurements from a 15-km-long aircraft transect with point-based flux measurements collected on the ground. When such sources of uncertainty are not documented and accounted for in the metadata for any given data set, biases can be introduced that affect the outcome of data analyses and the conclusions drawn from them. As a general rule, uncertainty and sensitivity analyses should routinely be accomplished for the outputs of all models using the same data sets.

The committee recommends that the metadata for each data set explicitly describe all preliminary processing associated with that data set, along with its underlying scientific purpose and its effects on the suitability of the data for various purposes. Further, the metadata also should describe and quantify to the extent feasible the statistical uncertainty resulting from each processing step. Planning for studies that involve interfacing should explicitly consider the effects of preliminary processing on the utility of the resultant integrated data set(s).

Metadata issues also are discussed in more detail later in this chapter.

Data Volume

The sheer quantity of data projected for global change studies can pose significant challenges for nearly every aspect of the data storage, retrieval, and analytical systems currently available. As summarized by Townshend and Rasool (1993), these drastically increased volumes stem from a variety of sources:

- A greater number of different sensing systems.
- An increase in the number of spectral bands and frequencies per sensor system.
- Improved sensor sensitivity.
- Cumulative increases in data volume as the historical record grows and sensor technology continues to advance.
- The proliferation of derived data sets to meet different needs.
- The creation of regional- and global-scale data sets from preexisting, fragmented data.

In many cases, the projected volume of data from new instruments and programs is orders of magnitude higher than that currently produced by existing programs. This can make it impossible or impractical to continue using traditional data management and data interfacing methods. For example, the staff at the Carbon Dioxide Information Analysis Center (CDIAC) at Oak Ridge National Laboratory are concerned that their existing labor-intensive data cleanup and interfacing methods will not be suited to the demands of the new Atmospheric Radiation Measurements (ARM) program. This DOE program will produce approximately 1,000 gigabytes (1 terabyte) of data per year, an amount significantly larger than the 5 gigabytes of data archived by CDIAC. Up to now, two of CDIAC's most popular data sets, Keeling's atmospheric carbon dioxide concentrations from Mauna Loa and Marland's carbon dioxide emission estimates from fossil fuel burning, are only 0.03 and 75.6 megabytes in size, respectively.

Larger data volumes also can require changes in the relationship between data sources and users. For instance, the FIFE Information System staff considered it unreasonable to respond to open-ended investigator requests such as, "Send me all your level-1 AVHRR-LAC (Local Area Coverage) data," which totaled 1.5 gigabytes on ninety 6,250-bpi 9-track tapes. Instead, user support staff worked with users to refine specific data requests for actual research requirements.

A related barrier stems from the need to provide the research community with information about and ready access to an ever-increasing volume of data. These volumes threaten to overwhelm some of the existing methods for data storage, archival, and retrieval.

The committee recommends that all proposed data management and interfacing methods be weighed carefully in terms of their ability to deal with large volumes of data. Assumptions that existing methods will continue to be suitable should be treated with caution.

Overcoming Data Incompatibilities

Data interfacing is often confounded by differences in the conventions that structure day-to-day practice. Each discipline has its own set of conventions, or language, which cannot easily be forced into a common terminology even when the same quantities are involved. For example, cartographers use "small scale" to refer to a very large area mapped without much detail, while other scientists use the same term to refer simply to a small area. Conversely, cartographers use "large scale" to refer to small areas mapped in great detail. Other scientists use this term to refer to extensive or large areas.

The FIFE and NAPAP studies provide a rich variety of examples of how seemingly innocuous data characteristics can bedevil data interfacing efforts. In the FIFE study, the Information System staff found that the same symbols or terms were used by different disciplines to refer to widely different quantities. An analogous problem arose from the fact that separate groups measured the same variables, but called them by such different names that it would be difficult to combine them without prior knowledge of their respective naming conventions. As another example, all disciplinary groups in the FIFE study measured time, but in ways that made it difficult to combine their respective data. The surface biology group preferred wall clock time for field data collection, while more globally oriented investigators used Greenwich Mean Time as a more universal standard to link satellite and aircraft data collection. Further, because time of day is not traditionally noted with soil moisture measurements, it was difficult to use the soil moisture data with other data, such as those for surface fluxes, with higher temporal resolution. This incompatibility was problematic because diurnal variations in soil moisture can have significant effects on energy and moisture balance calculations. As yet another example, investigators working at a local scale preferred Universal Transverse Mercator coordinates, while latitude and longitude were needed to track satellite and aircraft operations and to link solar position and time in a scientifically consistent way. The surface flux group was focused on circulation modeling with grid cells several kilometers on a side. That group was therefore satisfied with fairly coarse digital elevation data (e.g., 30-m horizontal resolution). In contrast, the hydrological investigators wanted watershed descriptions at 1-m resolution.

The FIFE study encountered even more severe problems when attempting to incorporate needed data from outside sources. The historical precipitation data, extending back to the 1850s, had no associated locational or methodological documentation. Similarly, historical vegetation cover maps were available, but contained only the names of vegetation classes and not full descriptions of the species present. County soil maps, obtained from the Soil Conservation Service, were difficult to reconcile across county boundaries because different, and poorly documented, classification schemes were used in different counties. Likewise, the aquatic component of NAPAP relied on soil maps and geological data collected by different state agencies; in a number of cases, different data conventions mandated additional field studies to make the data sets compatible. Poor documentation, differences in sampling methods, and inadequate site data for fisheries also hampered NAPAP efforts to integrate these historical data sets from state agencies with water chemistry data collected by the NAPAP agencies. These difficulties, and the high cost of collecting new fisheries data, severely curtailed assessments of acid rain impacts on biotic resources.

Problems of data compatibility due to different methods, different definitions, or a lack of coordination among parties are exacerbated for international data sets. For instance, in the Sahel study, historical crop yield, a crucial biological endpoint, was often poorly defined and monitored in different ways among participating countries. The lack of suitable and consistent ground truthing (i.e., verifying remotely sensed observations by comparing them with reliable in situ observations) of yield data was especially problematic.

These differences between ecological and geophysical studies, and the data they produce, cause significant problems for data interfacing efforts. Creating synoptic ecological data sets that match the broader coverage of geophysical data is time-consuming, costly, and technically demanding. Gathering ecological data over large areas and times is often labor-intensive and difficult. It may be necessary to combine data from several ecological studies, which in turn usually requires extensive data standardization and cross-checking among ecological data sets that were originally collected independently. Some problems, such as differences in spatial registration of mapped data, are potentially resolvable. Others, such as differences from study to study in measurement methods or class limits of key variables, may not be. Resolving these and other problems stemming from differences between geophysical and ecological data requires in-depth understanding of the data's characteristics and of the underlying scientific assumptions they reflect.

Each of the challenges described above gives rise to a corresponding requirement for successfully interfacing ecological and geophysical data. The methods for fulfilling these requirements will differ somewhat depending on whether interfacing is dealing with historical data from disparate studies or is envisioned for future studies that can be planned as integrated wholes. For historical data, requirements may have to be met by "retrofitting" certain aspects of the data. For future studies, challenges can be dealt with in the planning process, thereby avoiding some of the problems inherent in historical data. However, it is very unlikely that future interfacing efforts will occur only in the context of studies planned as fully integrated wholes. Instead, it is most likely that a great deal of interfacing will continue to involve data drawn from separate studies that were not originally envisioned as directly related. In either case, the degree to which these requirements are met will largely determine the success of interfacing efforts.

The committee recommends that efforts to establish data standards focus on a key subset of common parameters whose standardization would most facilitate data interfacing. Where possible, such standardization should be addressed in the initial planning and design phases of interdisciplinary research. Early attention to integrative modeling can help identify key incompatibilities. The data requirements, data characteristics and quality, and scales of measurement and sampling should be well defined at the outset.

In addition, the committee recommends that agencies that perform or support environmental research and assessment generally, and global change research particularly, identify and define key ecological data sets that do not exist but are important to their mission. A careful review should be made of options for finding, rescuing, or creating these crucial data, and funding should be set aside to implement the most feasible option(s).

ADDRESSING BARRIERS DERIVING FROM USERS' NEEDS

There is an extremely wide range of users from different countries, agencies, and scientific disciplines who could potentially use data interfacing in their work. However, in order for them to even conceive of an interfacing effort, they must first know what data are available in their particular discipline and perhaps even in the research community at large. Thus, a major threshold challenge is for data managers to keep a large, diverse, and geographically dispersed user community informed and up to date about data types, characteristics, locations, and retrieval methods, as well as about changes in these. Inevitably, there will be as many ways of perceiving and thinking about the data as there are users. Different users will want to subdivide the same data differently, will be interested in different temporal and spatial scales, will use dissimilar derived variables, and will focus on separate processes (instances of such differences were described above in the section "Addressing Barriers Deriving from the Data"). Data interfacing systems and procedures therefore must have the ability to represent the data in different ways that conform to users' mindsets and research needs. (See also the subsection "Interoperability" in the section "Addressing Barriers Deriving from Information System Considerations," below.) Further, because analytical approaches and methods in climate change research are not standardized, users need the ability to interface data with their own desired analytical tools. Even in the FIFE program, which was designed from its inception as a set of integrated investigations, researchers preferred whenever possible to analyze data on their own systems and with their accustomed tools.

Multidisciplinary problems demand that users retrieve and interface data from studies in which they were not directly involved. For example, in NAPAP, data on aquatic biota in studies of acid rain effects were collected by one group of scientists, while data on precipitation chemistry were collected by another. Similarly, in the FIFE program, modelers used atmospheric as well as point-source vegetation data to develop a picture of how processes at different scales interact. In these multidisciplinary programs, data were collected with the intention of integrating them later with other data types. In contrast, much of the data envisioned for use in global change studies were not originally collected with specific interfacing applications in mind. This situation requires that data be exceptionally well documented in order to be usable by scientists who were not directly involved in their original collection and validation. Working scientists readily recognize the difficulties presented by poorly documented data in their own field. Such difficulties only increase when data being interfaced cut across disciplinary boundaries. In fact, the committee heard researchers in every case study describe the confusion, inefficiency, and technical dead ends that can result from poor documentation. As a result of experiences such as these, many recent reports on data management related to global change research have emphasized the central importance of thorough and readily accessible metadata (e.g., NRC, 1991; OSTP, 1991; CEES, 1992). Providing such metadata can have implications for the design of hardware and software systems to support interfacing (see the subsection "Complex Metadata" in the section "Addressing Barriers Deriving from Information System Considerations," below).

Most of the case studies the committee examined operated on the assumption that data were a common resource intended to be used in different ways by different researchers. Indeed, this is a fundamental tenet of the global change research effort, where key data sets, such as those from remote sensing, are made available to the research community at large. When the same data are used in more than one way by different investigators, however, multiple and incompatible derivative versions of the original data set are produced. This is because researchers commonly process, subdivide, adjust, summarize, and correct working versions of acquired data sets based on their needs and their interactions with the principal investigator(s). In fact, where different researchers independently update or correct acquired data, a single, identifiable authoritative version may cease to exist. This happens when no one has the responsibility to ensure that all corrections and updates are collected and embodied in a validated "source" data set. Thus, for instance, an important aspect of CDIAC's role is to collect, document, and incorporate feedback from users into its data packages, such as the one for global production of carbon dioxide. This ensures that the user community can readily identify and obtain authoritative versions of key data sets. In contrast, analysts in the Sahel study, under pressure to produce timely crop estimates, made many undocumented corrections, adjustments, and derivations of original data. Therefore, even though some of the derived data sets that resulted from the Sahel study have been archived by NOAA, it is no longer clear how they were created. Thus, is it extremely difficult to use these data for other purposes or to retrace the data path to an intermediate point.

Global change research can be an extremely fluid activity. Scientists cannot always define exactly what data they are interested in, how they will interface them, or how they wish to analyze them. Even relatively well-defined programs involve a large component of exploratory data analysis. Researchers experiment with data, mixing and matching different data types to explore a variety of kinds of relationships among data and underlying processes. In addition, the results of one particular line of investigation can lead to another that was not in the original program plan (see Box 8.4). As a result, data interfacing systems and procedures must be exceptionally flexible and responsive to ad hoc retrieval requests, while data formats and transfer mechanisms must be standardized or otherwise made easily convertible (see the subsection "Interoperability" below).

Data interfacing systems and procedures must be adaptable as well over the somewhat longer term. The data, analytical methods, models, motivating questions, and related hardware and software all are changing and evolving rapidly. The committee heard unequivocal statements from participants in every case study that a critically important condition for successful data management and data integration is the direct and

> **BOX 8.4**
> **Users' Needs Can Change Unpredictably**
>
> The FIFE study provides a good example of how users, their scientific interests, and their needs for data all shift over time in ways that cannot necessarily be predicted. The overall study was based on a conceptual framework that guided the data collection and helped organize how the data would be used in atmospheric circulation models. However, FIFE was designed as an active experiment. It was not a well-defined monitoring program or a platform with a limited set of instruments. Investigators proposed measurements that had not been done before, built new instruments and modified existing ones, and undertook novel data collection procedures and protocols. When "quick-look" analyses identified aspects of the program that were not working, these were modified. The actual relationships among the measured variables, and how these were affected by sensor characteristics and radiative transfer processes in the vegetation canopy and in the atmosphere, were unknown at the beginning of the program. These became clear only over time as investigators began analyzing their data. As analyses progressed, investigators' interest in and need for data from other aspects of the study increased. In addition, the investigators themselves changed throughout the course of the experiment as graduate students and postdoctoral researchers moved on and principal investigators changed institutions.
>
> The need for certain data products, such as satellite image data processed to uniform formats, was anticipated in the planning. However, some anticipated needs never materialized. It was expected that many investigators would request AVHRR-GAC (Global Area Coverage at 4- to 8-km resolution) data, and substantial effort was expended to develop such data. However, the subsequent availability of an equivalent AVHRR-LAC (Local Area Coverage) data set at higher, 1-km, resolution absorbed all the anticipated demand. To date, there has not been one request for the GAC data set.
>
> The FIFE data also have been used in many ways not originally conceived. For example, they have been used as a source of test data for developing satellite processing algorithms. They also have been proposed as a test data set for producing data compression algorithms and data visualization tools.

continuing involvement of the scientific end users. Ensuring such involvement has been a consistent recommendation as well of recent reports on data management aspects of global change research (e.g., NRC, 1991; OSTP, 1991; CEES, 1992).

The committee urges project scientists and data managers to adopt the view that one of their primary responsibilities is the creation of long-lasting data and information resources for the broad research community. Data management systems and practices, particularly the development of metadata, should be designed to balance the needs of this larger user community with those of project scientists.

ADDRESSING BARRIERS DERIVING FROM ORGANIZATIONAL INTERACTIONS

Because few existing organizational arrangements reflect the multidisciplinary perspective that motivates the interfacing of geophysical and ecological data, any such interfacing will necessarily have to be implemented across a range of organizations. This in turn will involve making adjustments or accommodations to those aspects of organizations that affect the behavior of individuals, groups, and agencies. These include such things as management structures, organizational history and memory, norms of acceptable behavior, rewards and sanctions (both implicit and explicit), agency missions, and budgets. The committee heard in every case study that such organizational interactions were central to the success or failure of data management and data interfacing efforts. As a result, the committee concludes that organizational and technical considerations interact strongly and should be given equal weight in the design and development of data interfacing systems.

Rewards, Priorities, and Ingrained Attitudes

The reward system among scientists in universities is based largely on the publication of peer-reviewed papers. Within this overall setting, the geophysical sciences have a longer and more established tradition of studies with multiple investigators and complex data sets. In contrast, the ecological sciences are much more oriented toward studies with a single principal investigator. For example, in tenure decisions, many academic ecology departments give much greater weight to sole-author and first-author publications than to those with larger numbers of authors. While the size of the research group thus varies from field to field, the chief priority remains to publish original research in the peer-reviewed literature. Therefore the primary objective of most scientists is to gather, process, analyze, and publish "their" data; only then do they respond to outside demands. But successful data interfacing requires that scientists participate more fully in outside activities (e.g., preparation of documentation, exhaustive quality control, and data submissions to outside agencies) for which they are not directly rewarded.

An analogous reward system influences the behavior of scientists and managers in agencies. Here, rewards stem from furthering the agency mission, as well as from publishing. In most cases, agency missions, and those of the departments or divisions within them, have been narrowly defined to focus on a particular activity, resource, or problem. This approach reflects the classic bureaucratic organizational model in which problems are addressed by compartmentalization, division of labor, and well-defined procedures (Parsons, 1947). In most cases, the additional activities needed to support data interfacing (e.g., preparing documentation, responding to data requests, participating in standardization efforts) do not fit within traditional agency missions. Not only are staff typically not rewarded for such activities, in some circumstances they may be chastised for threatening an agency's highly prized autonomy. Thus, data interfacing efforts must overcome various versions of the "It's not in my job description" problem.

At an extreme, such inappropriate reward systems and the mindset they engender can result in the partial or complete failure of data interfacing efforts. For example, the documentation submitted to the Information System staff in the FIFE study was of variable quality, much of it insufficient to enable other researchers to make use of the data. Some data sets were fully described, but many others were accompanied only by a list of files, variables, and formats. The Information System staff found that many investigators were reluctant to devote the time and effort needed to prepare or review documentation, even when draft documents were prepared by the Information System staff. While most of these problems were overcome by persistence, the inability to develop complete documentation rendered two data sets—micrometeorological data from the Army Corps of Engineers and GOES data processed by Scripps Institution of Oceanography—essentially inaccessible and unusable. Because of problems such as these, the FIFE Information System staff identified the preparation of documentation as the single most demanding and frustrating element of their data interfacing experience.

Overcoming these obstacles sometimes requires accommodating or mimicking the existing reward system. For example, CDIAC often negotiates with data sources to allow them time to analyze their data and publish their results before submitting data for wider distribution. Consequently, the center sometimes waits for up to 2 years to receive targeted data sets. By listing data sources as primary authors of data packages once they are published and encouraging users of the data to cite data sources in their publications, the Center tries to use the academic reward system as a motivation for data sources to participate actively in the preparation of the data packages. Whatever the means used, successful data interfacing depends on surmounting the ingrained mindsets and priorities fostered in research scientists by the existing organizational reward system.

The committee recommends that professional societies, research institutions, and funding and management agencies reevaluate their reward systems in order to give deserved peer recognition to scientists and data managers for their contributions to interdisciplinary research. Granting and funding agencies, as well as program managers and uni-

versity administrators, should provide tangible incentives to motivate scientists to participate actively in data management and data interfacing activities. Such incentives should extend to favorable consideration of those activities in performance review, including treating the production of value-added data sets as analogous to scientific publications.

Also, because organizational missions and reward systems inherently reflect a larger policy context, relevant policy issues should be included in the planning for interdisciplinary research. This should be accomplished in part through open communication between project scientists and appropriate policymakers that continues throughout the life of the project. Such communication will help provide a basis for developing cooperative arrangements between collaborating institutions that will provide strong incentives for and reduce barriers to sharing data.

In addition to these academic and organizational reward systems, barriers to data interfacing stem from ingrained attitudes about the relative professional status of different groups. In every case study the committee found that interfacing geophysical and ecological data typically required information management skills that were beyond the capabilities of the average scientist. The lack of these skills is particularly acute in the ecological sciences, where, because studies are generally finer-scale and shorter-term, scientists do not need to confront complex data management issues. Ecologists are more likely to use software packages that combine statistical analysis capabilities with rudimentary data management features. These packages do not typically have the ability to handle complex relational databases or geographically based data, and, if they do, ecologists are less likely than geophysicists to use them. Relatively straightforward data management approaches are usually sufficient for the characteristic ecological study, where data volumes rarely exceed a few hundred megabytes, but not for the gigabyte- and terabyte-sized data sets typical of geophysical studies.

Despite the geophysical science community's greater familiarity with sophisticated hardware and software, however, both communities are relatively unschooled in the information management concepts and system design skills required for successful data interfacing. As a result, the most successful data interfacing efforts in the case studies were those where information management professionals were an integral part of the research team (Kanciruk and Farrell, 1989). Involving such individuals in the effort in turn required surmounting subtle ingrained attitudes that can frustrate productive cooperation between geophysicists and ecologists on the one hand, and information management specialists on the other. Each group is proficient in its own field and relatively naive in the other's area of expertise. All too often, therefore, neither group affords the other the status of equals. There is a tendency among scientists to view information management staff as "technicians" fulfilling a support function that does not warrant equivalent status. Conversely, there is an inclination among information management specialists to view scientists as merely "users," a term that can have pejorative connotations. Successful broad-scale interfacing efforts require strategies that break down such ingrained attitudes and equalize the perceived status between the two groups.

In the NAPAP study, conflicts developed between the data managers and the scientific team. On the one hand, the scientific team felt that the data management team should provide a workable database from which the scientific team could do the actual data analysis and interpretation. On the other hand, the data management team considered themselves scientists as well as data managers and wanted the opportunity to interpret the data also. In the best of all worlds, the data management team and the scientific team should be combined to constitute a single team that can and should jointly prepare and publish interpretative reports.

The committee recommends that in order to help ameliorate some of these difficulties, research universities include courses in their curricula that provide environmental scientists with an in-depth understanding of the rationale for and principles of sound data management. Program managers and data managers, in their interactions with and training of environmental scientists, should emphasize how state-of-the-practice data management can provide immediate and long-lasting benefits to scientists, particularly those engaged in interdisciplinary research. At the same time, data managers need to be a part of the conceptual team from the beginning of a project and have equal status with principal investigators.

Data Ownership and Cooperation

The reward systems and attitudes described above contribute to a distinct sense of possessiveness about data. This is especially acute prior to publication when the fear of being "scooped" is highest. As noted in the previous section, CDIAC's data sources are sometimes unwilling to submit data for wider distribution until they have first published. Existing funding mechanisms also contribute to this sense of ownership of individual data sets. Even in the FIFE study, which was intended from the outset to involve a large amount of data interfacing across studies and data types, contracts were awarded to individual investigators or groups whose proposals were selected. In the Long-Term Ecological Research (LTER) study, data sharing among scientists increased in direct proportion to the amount of use the principal investigators were making of other researchers' data. As a result, scientists have a built-in motivation to think of data as belonging primarily to their particular project and only secondarily to larger data interfacing efforts that cut across projects.

Analogous concerns about data ownership can be even more intractable at the level of agencies or organizations. Here, data are usually viewed as an important organizational or economic resource, and efforts to make them more widely available to "outsiders" can be viewed with suspicion. In some cases, this suspicion stems from legitimate concerns that sensitive information will be misinterpreted or misused, and in others from a fear of being made to look bad. In other instances, organizations request payment or a fee as a return on the investment needed to gather the data in the first place. This request is often made for geophysical data that might be useful for such purposes as mineral prospecting or crop management. There also may be legal or ethical constraints on making data available when they contain proprietary or privileged information, or where their release could violate privacy rights.

The issue of data ownership arose in every case study the committee examined. Data interfacing succeeded only when explicit steps were taken to develop organizational mechanisms that produced countervailing motivations. For instance, both the California Cooperative Oceanic Fisheries Investigation (CalCOFI) and the FIFE programs were designed so that individual scientists need to have access to other kinds of data in order to address the program's core questions. However, data ownership problems cannot always be resolved in favor of complete openness and access to data. In the case of CDIAC's global inventory of carbon dioxide emissions, the Center had to accommodate certain politically motivated restrictions on the ways in which country-by-country population data could be reported.

Considerations of institutional ownership also can prevent data integration. For example, NAPAP relied on historical fisheries data from state agencies and found some cases in which well-documented and well-managed data sets were restricted from outside use.

The committee recommends that in order to encourage interdisciplinary research and to make data available as quickly as possible to all researchers, specific guidelines be established for when and under what conditions data will be made available to users other than those who collected them. Such guidelines are particularly important when data collectors, data managers, and other users are in different organizations. In addition, adequate rewards should be established by the funders of research and publishers to motivate principal investigators to place all data in the public domain.

Managing Organizational Change

Responses to these kinds of organizational barriers, as well as to those deriving from the data, from users' needs, and from system considerations, will necessarily be implemented by managers of and participants in data interfacing efforts. To succeed, they will need the ability to achieve three kinds of changes in existing organizational interactions. First, they must establish relationships and operational procedures across as opposed to strictly within organizations. Second, they must establish reward structures and other mechanisms to induce organizations and the individuals within them to change their behavior. Third, they must establish new functions that support data interfacing, but that are currently beyond the responsibilities of any individual organization.

These changes are analogous to those being implemented across a wide range of industries in the private sector. In these instances, managers are attempting to integrate previously separate activities or functions in order to achieve greater quality, productivity, and flexibility. Both the process of organizational change in industry and its substance are similar to the organizational changes required in data interfacing. A great deal of the learning taking place in these private sector organizations is thus directly relevant to the data interfacing context. There is a large and growing literature, from both corporate and academic perspectives, that presents and analyzes this experience. A thorough review is beyond the scope of this report, but managers of data interfacing efforts would benefit from reading some or all of the following: Davenport (1993), Freudenberg (1992), Katzenbach and Smith (1993), Lincoln (1985), Reason (1990), and Weick (1985).

As might be expected, given the wide range of organizational sizes, structures, histories, and cultures, there is no single formula for successfully resolving the challenges deriving from organizational interactions related to data interfacing. The particulars of successful solutions depend on the context of individual situations. However, several rules of thumb can be inferred from past experience, both within the case studies and in industry at large. Successful management of data interfacing efforts must be based on collaboration and flexibility, and specific attention must be paid to reward structures and motivation. In addition, management must foster productive teamwork between geophysical and ecological scientists, as well as between scientists and information management specialists.

The committee recommends that in the planning of any interdisciplinary research program, as much consideration be given to organizational and institutional issues as to technical issues. Every effort should be made to minimize the likelihood of misunderstanding, conflicts,

and rivalries by establishing interorganizational relationships and procedures, creating effective reward structures, and creating new functions that explicitly support data interfacing.

The committee also recommends that the agencies involved in supporting and carrying out interdisciplinary research investigate the possibility of establishing one or more ecosystem data and information analysis centers to facilitate the exchange of data and access to data, help improve and maintain the quality of valuable data sets, and provide value-added services. A model for such a center is the Carbon Dioxide Information Analysis Center (CDIAC) at Oak Ridge National Laboratory. In addition, it would be wise to look closely at the potential synergism between any new ecosystem data and information analysis center and all other existing environmental data centers.

ADDRESSING BARRIERS DERIVING FROM INFORMATION SYSTEM CONSIDERATIONS

Actual interfacing activities will be carried out primarily by means of computerized hardware and software systems, and interfacing capabilities will depend in large part on their characteristics. The range of challenges deriving from users' needs, from the organizations in which they work, and from the data themselves complicates the design, implementation, operation, and maintenance of such systems.

The wide diversity of studies, data types, and researchers is matched by a similar diversity of preferred hardware and software. It is unrealistic to expect that the research community at large, or even a sizable segment of it, could be required or persuaded to use common hardware and software. This inescapable variety is compounded by the rapid rate at which hardware, software, and system concepts are evolving and the fact that they are unlikely to stabilize soon. In this environment of rapidly shifting requirements, preferences, and capabilities, interfacing approaches, and the tools that implement them, should ideally be as independent as possible of specific platforms or systems. At present, however, the lack of true interoperability among different hardware and software components makes it impossible to completely achieve this ideal.

The demands of interfacing will stretch current concepts of system design and development. They will do this in at least four specific ways. First, systems must be developed in the absence of clear-cut, detailed, and stable users' requirements. Second, systems must more fully achieve the kind of interoperability that will give users flexible access to widely dispersed data. Third, metadata must efficiently describe the location and characteristics of a proliferating number of ever-larger data sets. This requires that metadata become more closely integrated with the data they describe. Fourth, users must have the ability to backtrack along the data path when derived data do not meet the needs of a particular interfacing scenario. All four of these challenges, described more fully below, are compounded by the massive amounts of data involved in global change and other complex environmental research programs.

Fuzzy and Shifting Requirements

Some aspects of users' needs in global change research help to create exceptional challenges for system designers. In a typical system design scenario, clear definitions of users' requirements provide the basis for the detailed design of the hardware/software system. However, scientists engaged in interdisciplinary environmental research cannot specifically define all the kinds of interfacing they might wish to do in the present, nor how these desires might change in the future. Thus, while users are usually able to define a broad vision (e.g., maximize accessibility, flexibility, and adaptability), they cannot specify detailed and concrete requirements for interfacing (e.g., take exactly these data and do precisely these, and only these, operations on them). The research setting is thus fundamentally different from the setting of many large business systems (e.g., airline reservations and automated teller machine networks) for which requirements can be completely specified. In this kind of environment, as exemplified in the FIFE study described above in Box 8.4, users and information management specialists must interact closely and more continuously in order to respond to shifting needs (Figure 8.2) (Desmedt, 1994a,b).

Incompletely defined and constantly changing users' requirements mean that systems that support interfacing must be designed for flexibility in the present and adaptability in the future. Because such systems can never be actually finished, a fundamental shift in mindset is required on the part of system designers. In fact, striving to freeze users' requirements in order to deliver a stable, completed system can be counterproductive. For example, initial data management and interfacing efforts in the FIFE program were based on a traditional engineering approach in which system designers who lacked the requisite scientific expertise gathered input from users, established a set of users' requirements, and then attempted to develop and deliver a completed system without further interaction with the intended users. Not surprisingly, this design approach failed and had to be replaced with one based more directly on active and ongoing interaction between the information management staff and the research scientists.

In the Aquatic Processes and Effects portion of NAPAP, atmospheric and geochemical data collection and management activities were well-defined, well-focused, and consistent through the duration of the project.

Sequential (closed)

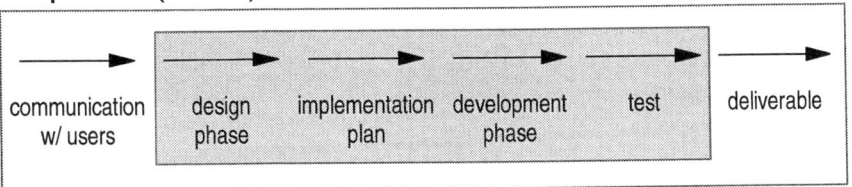

Parallel (collaborative)

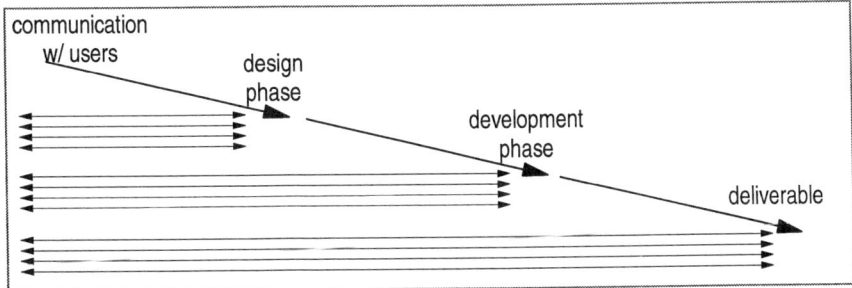

FIGURE 8.2 Schematic representation of the difference between classical sequential system design and a more flexible collaborative approach. In the sequential design, information management specialists interact with users only at the beginning and end of the process. In the collaborative process, users and information management specialists work in parallel to contribute their knowledge and insight to the design as it develops (adapted from Strebel et al., 1990).

Thus, data integration went smoothly based on watershed models, and well-documented databases were made available to a wide user population in a timely fashion. In contrast, the technical scope for fisheries assessment was not clearly defined in the early stages of program development and was inadequately represented in new field studies later on in the project. Uncertain and shifting requirements with regard to biological data led to an overreliance on poorly documented historical data, often collected and managed by state agencies. As a result, data integration was difficult and compromised technical rigor in assessing acid rain impacts to fish populations. Also, efforts to maintain and disseminate an integrated fisheries database were abandoned.

The committee recommends that hardware/software system development efforts be based on a model that includes ongoing interaction with users as an integral part of the design process. In addition, system designers should work from the assumption that systems will never be

finished, but will continue to evolve along with the data collected and users' needs. Designers therefore should use, to the greatest extent possible, modern database development approaches such as rapid prototyping, modular systems design, and object-oriented programming, which enhance system adaptability.

Interoperability

Poorly defined and changing requirements, along with many of the other challenges to data interfacing described above, could easily be satisfied by true interoperability. Interoperability, as defined here, is the ability to readily connect different databases on separate hardware/software systems and perform data retrieval, analyses, and other applications without regard to the boundaries between the systems. Such seamless interoperability is the basis for the concept of "virtual databases" in which users have access to a wide variety of data, regardless of their location (NRC, 1991).

Despite progress toward this goal, anyone who has attempted to functionally connect different hardware and software knows that this can be tremendously demanding, complex, and frustrating. In general, these difficulties derive from two sources. First are the differences in hardware- and software-specific issues such as digital communication protocols and ways of structuring and indexing databases. Surmounting these problems requires the development of common digital interface standards, either by the user community or through cooperative efforts of hardware and software vendors. In either case, the task can be arduous and time consuming. Second are the semantic differences between data from disparate databases. These result from the data themselves, as discussed earlier in this chapter, and include, for example, incompatible scientific naming conventions and fundamental differences in spatial and temporal scale. This second kind of barrier to interoperability can be overcome only by standardization of the key parameters that allow linkages to be created among different data sources. In the FIFE study, for example, the wide assortment of methods for recording time of sampling made it difficult to interface measurements taken at the same time.

These two barriers to interoperability must be approached and overcome in the context of a third and related barrier. This is the problem of providing users with flexible access to a diversity of data sources that will be subdivided and recombined in a variety of ways that cannot always be predicted. Although it is advancing rapidly, current technology cannot yet provide users with the flexible access to, retrieval from, and combination of data from widespread sources that they desire.

The committee recommends that program managers, project scien-

tists, and data managers review the interoperability of their hardware, software, and data management technologies to facilitate locating, retrieving, and working with data across several disciplines. However, this effort should be accompanied by parallel attempts to resolve inherent incompatibilities among data types that can thwart interfacing even when state-of-the-art hardware and software systems are seamlessly connected.

Complex Metadata

Many discussions of the data management issues related to global change research have emphasized the importance of accurate and complete metadata (e.g., NRC, 1991; OSTP, 1991; CEES, 1992). Box 8.5 expands on the committee's definition of metadata, which is provided in the Executive Summary. Practicing scientists understand that no data set is perfectly consistent, free of peculiarities, immune to errors, or necessarily suitable for all analytical approaches. They therefore depend on metadata to describe the data, suggest limitations on its use, and warn of potential pitfalls. Metadata become critically important in the interfacing context as scientists work with data that are outside their area of primary expertise and as they use data in ways not originally intended. At present, most metadata are typically contained in a document separate from the data themselves. For example, CDIAC provides hard-copy metadata with each of its data packages, and the NASA Global Change Master Directory describes available data sets in on-line metadata. In addition, certain kinds of information about individual data points can be embedded in the data set as codes or flags that, for example, identify data points of questionable quality.

This approach has proved satisfactory for relatively small data sets. However, the committee found widespread concern among data management specialists that the proliferating and ever-larger data sets used in global change research would make this approach unworkable, especially when interfacing different data types. Three principal and related concerns were voiced. First, the volume of some kinds of metadata is likely to increase along with the size of the data sets they describe. This is particularly true of the portion of metadata that refers to the peculiarities of individual data points or groups of data points. It may be unrealistic to expect researchers to be able to effectively assimilate this amount of information. Second, while some of this more specific metadata can be encoded as fields or flags associated with the data themselves, these codes and flags will not be relevant to all interfacing situations. This is because a particular data characteristic will be a greater or a lesser problem depending on the circumstances (see Box 8.6). As a result, researchers who

> **BOX 8.5**
> **Metadata: The Key to the Lock**
>
> Metadata document or describe all the facts, circumstances, and conditions associated with the actual data themselves. In most cases, metadata are the key needed for scientists other than the original investigator(s) to unlock the information contained in the data. This is because they provide insight into not only the raw characteristics of the data, but also constraints on their use and limits on their interpretation. Metadata are thus essential to the process of drawing scientifically defensible conclusions from the data.
> The essential components of metadata differ from project to project and data type to data type. However, the key elements include at a minimum those listed here. In addition, the committee found that the most thorough and useful metadata results from the combined efforts of principal investigators, information management specialists, and other potential users not directly involved in collecting the data. Principal investigators are intimately familiar with the data and their quirks and peculiarities. Information management specialists are knowledgeable about ways in which the inherent structure of the data can affect their utility. Finally, other potential users will raise issues and propose applications of the data that would never have been thought of by the principal investigator.
> Key elements of metadata should include detailed description of at least the following (adapted from CENR, in press):
>
> - Identification of contributors.
> - Scope and purpose of the research program.
> - Data collection methods (field and laboratory), including description of instrumentation.
> - Sampling/measurement patterns in space and time.
> - Gaps or inconsistencies in sampling/measurement patterns.
> - Constraints imposed by measurement and processing methods.
> - Preliminary processing and derivation algorithms.
> - Quality assurance and control methods, including uncertainties in data or derived results.
> - Definitions and formats for each variable.
> - Quality control flags associated with the data.
> - Quirks and peculiarities of the data.
> - Limitations of the data.
> - Potential problems in specific kinds of applications.
> - Planned and actual applications of the data, including references to published papers.

are interfacing large data sets cannot necessarily depend on embedded codes and flags to automatically subdivide, transform, or otherwise operate on the data. Thus, while researchers cannot necessarily depend on automation to solve this data management problem, neither can they be expected to exhaustively examine these detailed metadata manually to evaluate their relevance. Third, there is a large and ever-increasing num-

> **BOX 8.6**
> **Metadata in an Interdisciplinary Context**
>
> A chronic problem with metadata is the reluctance of researchers to allow their data sets to be freely used by others. Good metadata, of course, are designed to allow this very thing to happen. Therefore, some mechanism needs to be established that encourages the free sharing of data.
> The H.J. Andrews Experimental Forest LTER study provides an interesting example of how a mechanism for sharing data across discipline boundaries developed on its own. The project started out as a series of related, but independent studies. As the study progressed, however, individual researchers began to realize that they needed data from other projects carried out at the site. As their need for other data increased, the individual scientists began to recognize that they had to make an effort to allow their data to be used by, and made useful to, their associates. This led to greater emphasis on adequately documenting their data sets.

ber of data sets potentially useful in interfacing applications. For researchers to make effective use of this variety, they require a means of identifying, evaluating, accessing, and retrieving relevant data. Metadata play a key role in providing the information necessary for these steps. At present, some of this information is available through a combination of publications (e.g., newsletters and catalogs from CDIAC and the National Geophysical Data Center) and on-line data directories. These avenues are suited to providing summary descriptions of available data sets. However, it will be a challenge to furnish researchers with an efficient source of information about all available relevant data without requiring them to search numerous catalogs and directories.

The traditional approach to metadata, in which they are considered as information separate from the data themselves, will not meet the challenges just described. The committee found agreement among a large segment of the data management specialists it consulted that a new conceptual model of metadata is required in which metadata are somehow integral to the data themselves. There was equally wide agreement, however, that no quick fixes to this problem are readily apparent.

The committee recommends that the production of detailed metadata be a mandatory requirement of every study whose data might be used for interdisciplinary research. Metadata should be treated with the seriousness of a peer-reviewed publication and should include, at a minimum, a description of the data themselves, the study design and data collection protocols, any quality control procedures, any preliminary processing, derivation, extrapolation, or estimation procedures, the use of professional judgment, quirks or peculiarities in the data, and an assessment of features of the data that would constrain their use for certain purposes.

Retracing Data Paths

As described above, all data sets reflect a particular set of users' needs and perspectives, which are not necessarily applicable to all situations. Interfacing ecological and geophysical data can therefore require that they be reformatted, resummarized, reclassified, or otherwise adjusted. For example, in the Sahel study, remote sensing and ground-based precipitation data were combined to develop an overall picture of rainfall throughout the study area. In many cases, this can involve backtracking down the data path to an earlier version of the data and then proceeding from there along an alternate path (Figure 8.3). The ability to perform this kind of backtracking requires that detailed information be available about the prior processing steps that were used to create the data set(s) being retrieved for interfacing. Sometimes this can be accomplished with thorough metadata. However, when a large user community is simultaneously using, updating, and modifying a considerable number of data sets (as in the FIFE study), stand-alone documentation is not adequate.

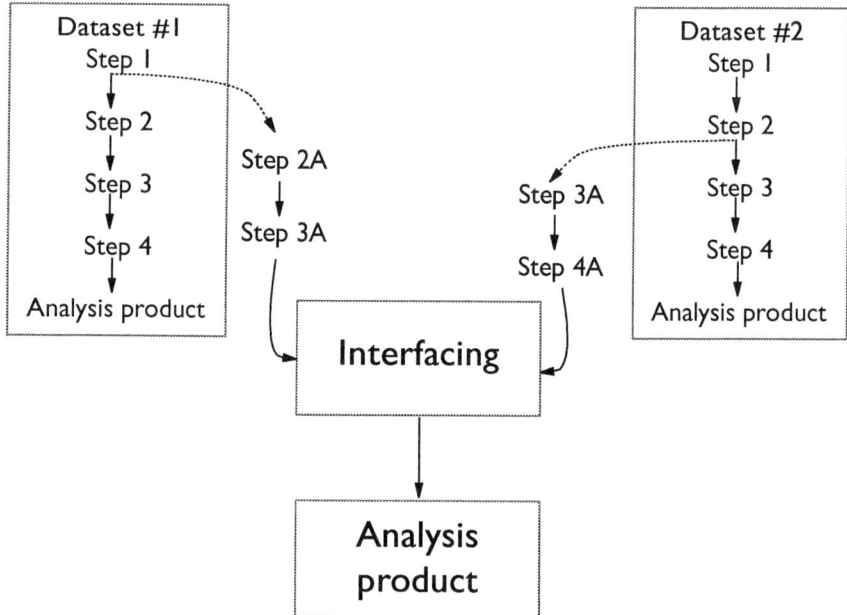

FIGURE 8.3 Schematic representation of how data interfacing can involve data processing steps that would not be needed if data were being analyzed independently. For each unique data set, this different data processing can require backtracking along the data to different points.

Instead, it is necessary to include this dynamic information in the database system itself by attaching it to each derived data set.

The committee recommends that metadata contain enough information to enable users who are not intimately familiar with the data to backtrack to earlier versions of the data so that they can perform their own processing or derivation as needed. Where stand-alone documentation is not adequate (for large and complex data sets or where multiple users are simultaneously updating and modifying data), data managers should investigate the feasibility of incorporating an audit trail into the data themselves.

Long-term Archiving of Data

An important concern is the stewardship of data sets throughout their life cycle, which for global change data extends over a minimum of decades to centuries and does not necessarily end when the primary users believe they no longer need the data. The committee concludes that far too many environmental research projects give insufficient attention, in either the planning or the implementation stage, to the long-term archiving of their data sets. Data from studies that contribute significantly to our understanding of components and processes of the Earth system must be preserved and made accessible for future potential users of the data. There is a need to create a mindset within the research community that valuable data must have a long-term life that extends far beyond the publication of the principal investigator's analyses.

In this regard, the committee found that there are no well-established and widely accepted protocols to assist scientists in deciding which data should be archived, in what formats they should be stored, and where and how they should be archived to maximize access for potential users. Further, in several cases the committee found little attention given to the long-term maintenance of data sets once they were archived. It is important to note, however, that there do not appear to be any insurmountable technical barriers to keeping all data collected in research projects, even data-intensive ones that involve high-resolution imagery, because advances in data storage and retrieval capabilities have kept pace with the ever-growing volumes of data in all fields of science. It is typical that the ensemble of all previous data in any scientific discipline is modest in volume compared to present and anticipated annual volumes. Therefore, the issue is not unmanageable volumes of data, rather it is the maintenance of the data sets in accessible, usable form over time that is the challenge for long-term retention.

These and other issues related to the long-term archiving of geophysical and environmental data are discussed in depth in another report,

Preserving Scientific Data on Our Physical Universe: A New Strategy for Archiving the Nation's Scientific Information Resources (NRC, 1995). However, the committee wishes to emphasize the following points regarding interdisciplinary data archiving:

The committee believes that, in general, the presumption in environmental research should be that "data worth collecting are worth saving." The committee therefore recommends that funding agencies consider stipulating that all research applicants include in their research plans well-conceived and adequately funded arrangements for data management and for the ultimate disposition of their data. While it is impossible to establish universal guidelines for funding, the committees's investigations suggest that setting aside 10 percent of the total project cost for data management would not be unreasonable. These cost estimates should include adequate funds for preparing thorough metadata that serve the needs of all potential users. In order for these requirements to be fully effective, however, the agencies must adequately support active archives and long-term data repositories.

Finally, the committee is concerned about the gaps in the existing system for long-term retention and maintenance of environmental data. The committee recommends that funding agencies provide guidelines that define the requirements for preparing data sets for long-term archiving. Educational and research institutions should be encouraged to incorporate strong data management and archival activities into every interdisciplinary project and should allocate sufficient funding to accomplish these functions. Professional recognition should be given to principal investigators and project data managers who perform these functions well.

TEN KEYS TO SUCCESS

The committee's investigations of the case studies and other related experience led it to identify 10 Keys to Success (Box 8.7), each of which incorporates both technical and cultural aspects. Keys 1 and 2 deal with the appropriate use of available information management technology. Keys 3, 4, and 5 describe design and management strategies. Keys 6, 7, and 8 refer to methods for accommodating the unavoidable realities of human behavior, motivation, and politics. Finally, keys 9 and 10 suggest ways of enhancing data interfacing by building a need for it into the structure of research programs. A discussion of the keys in terms of this grouping follows.

1. Be practical.
2. Use appropriate information technology.

> **BOX 8.7**
> **Ten Keys to Successful Data Interfacing**
>
> 1. Be practical.
> 2. Use appropriate information technology.
> 3. Start at the right scale.
> 4. Proceed incrementally.
> 5. Plan for and build on success.
> 6. Use a collaborative approach.
> 7. Account for human behavior and motivation.
> 8. Consider needs of participants as well as users.
> 9. Create common needs for data.
> 10. Build participation by demonstating the value of data interfacing.

The utopian ideal of a comprehensive technological solution for interfacing environmental data almost certainly will never become a reality. On the one hand, attempts at such solutions typically lead to disaster because they ignore the realities of users' shifting needs, diverse and evolving hardware and software systems, different personal motivations, and the generally complex organizational contexts of interdisciplinary research projects. On the other hand, failing to use appropriate, up-to-date information management technology can impede or even prevent data interfacing efforts. For example, the committee heard of numerous instances of the problems created by researchers' use of spreadsheets, rather than actual database software, for data management functions. Similarly, while checking data obtained from data sources, CDIAC staff found many errors that stemmed from a reluctance or inability to use more sophisticated programming languages to search for errors and discrepancies.

Balancing the constraints imposed by real-world practicality with a desire for the benefits of up-to-date information management technology is a difficult challenge. It requires close attention to users' needs and perceptions combined with the clever application of suitable technology. For example, both the FIFE Information System staff and the designers of the NASA Master Directory successfully applied a "least-common-denominator" principle for key elements of their systems. This helped achieve their goal of making these systems usable by the widest possible audience, even those with less than state-of-the-art hardware and software. However, while depending on a least- common-denominator user interface, the NASA Master Directory uses up-to-date networking technology to link widely separated databases. The CDIAC program provides another instance of this least-common-denominator approach. Its managers realized they could best fulfill their obligations as a key data resource by distributing their data packages via older, and more widely accessible, technology. As another strategy, the designers of the CalCOFI program accomplished their goal of developing a resilient long-term program by focusing on a few relatively simple parameters that would stand the test of time. The successful case studies chose technology that best accommodated constraints imposed by users while also furthering the project's fundamental goals. None of them used information management technology for its own sake.

3. Start at the right scale.
4. Proceed incrementally.
5. Plan for and build on success.

Successful interfacing efforts begin with discrete, well-bounded, and manageable pieces of the larger interfacing problem. This approach permits solutions to be developed and tested at appropriate scales as working relationships and the understanding of users' needs evolve through experience. Success at these functionally well-bounded scales builds confidence, credibility, and the desire of the participants to buy into the effort. Thus, information management specialists should help select initial interfacing applications that have a high probability of success. They also should design hardware and software systems that can evolve over time by expanding, adapting, and rearranging semi-independent modules. Users' needs demand systems that have the ability to evolve over time, rather than static and broad-scale "solutions" that are often obsolete by the time they are installed. The information managers in the FIFE program responded to the interfacing needs that arose naturally from the interactions of the project scientists (but see also keys 9 and 10 below). These incremental and nonthreatening responses helped build momentum and created a supportive environment for interfacing. Similarly, CDIAC initially focused on demonstrating success with a few fundamentally important data sets. It should be noted, however, that starting at too small or restricted a scale could lead to major problems in the future when it is realized, for example, that important functions have not been included in the system.

6. Use a collaborative approach.
7. Account for human behavior and motivation.
8. Consider needs of participants as well as users.

Both research scientists and information management specialists uniformly stress the importance of collaboration as the best way of dealing with the human element in data interfacing applications. This is particularly true of the process of hardware/software system design, where ongoing interaction between designers and users is vital to success. Collaboration also can help overcome potential political constraints to interfacing. Research scientists, used to relative autonomy, are not going to comply with data interfacing standards or procedures simply because someone tells them to. In addition, technology alone will not resolve such human issues. As Davenport et al. (1992) state, "No technology has yet been invented to convince unwilling [scientists] to share information or even to use it." The best means of ensuring active cooperation in facilitating data interfacing is to solicit information about users' priorities and concerns and to somehow account for these (e.g., Kirkpatrick, 1993). The best data interfacing solutions are those that make users' day-to-day lives easier and help them accomplish those things that are important to them. For example, CDIAC makes every effort to see that data sources are recognized for their contribution. It lists the data source as the primary author of each data package, provides a suggested bibliographic format for citing the data package, and encourages users to cite data packages as publications. The CalCOFI program has a history of collaborative decision making, in which project scientists participate equally in decisions about program direction and scope. As a result, the program has been able to adapt to changes in funding levels while maintaining the involvement of participating scientists.

The LTER network, with a large constituency of academic researchers (who are rewarded within the university system for individual research), has initiated a periodic research symposium as part of its ongoing efforts to promote data sharing and integration. Interactions among data managers and researchers at various sites help exploit historical data in new ways and encourage new proposals focused on intersite analyses. These activities are intended to lead to additional publications and funding of research—positive incentives for data sharing that NSF promotes with supplemental research funding.

It is important not to lose sight of the distinction between users and participants. Users are typically scientists who make use of a data interfacing system to retrieve information or data that further their own investigations. Participants, on the other hand, are in some way essential to the success of the data interfacing effort, but do not necessarily make use of the data themselves. Each data interfacing application involves a mixture of users and participants, both of whose needs must be taken into account.

9. **Create common needs for data.**
10. **Build participation by demonstrating the value of data interfacing.**

The potential conflict between data interfacing and individual users' needs can be resolved by designing programs so that users actually depend on successful data interfacing to meet their needs. For example, the fundamental scientific questions motivating the NAPAP and CalCOFI programs demanded that physical and biological scientists interface their data. Similarly, many of the projects within the FIFE program were designed so that researchers needed access to each others' data in order to meet their research goals. Tangible benefits can further motivate researchers to participate in data interfacing efforts (see also key 7 above). The FIFE managers judged that the earlier inclusion of an integrative modeling group would have stimulated much more interaction and data interfacing by making the value of such activities much more apparent. Such tangible demonstrations of interfacing's benefits are vitally important. Successful interfacing efforts require participants to change fundamental aspects of their attitudes and behavior. People are much more likely to make such changes if they can see actual examples of the promised payoffs.

Environmental research and monitoring projects are certain to become more complex and interdisciplinary in scope. By adequately planning for and anticipating the data interfacing issues raised in this report, researchers and data managers can help reduce the barriers and challenges that they will inevitably encounter.

REFERENCES

Baskin, Y. 1993. Ecologists put some life into models of a changing world. *Science* 259: 1694-1696.

Clark, W.C. 1985. Scales of climate impacts. *Climate Change* 7: 5-27.

Cole, K. 1985. Past rates of change, species richness, and a model of vegetational inertia in the Grand Canyon. *Am. Nat.* 125: 289-303.

Committee on Earth and Environmental Sciences (CEES). 1992. *The U.S. Global Change Data and Information Management Program Plan*. National Science Foundation, Washington, D.C.

Committee on Environment and Natural Resources (CENR). In press. *The U.S. Global Change Data and Information System Implementation Plan*. Joint Oceanographic Institutions, Inc., Washington, D.C.

Connell, J.H. 1985. The consequences of variation in initial settlement vs. post-settlement mortality in rocky intertidal communities. *J. Exp. Mar. Bio. Ecol.* 93: 11-45.

Davenport, T.H. 1993. *Process Innovation: Reengineering Work Through Information Technology*. Harvard Business School Press, Cambridge, Mass. 337 pp.

Davenport, T.H., R.G. Eccles, and L. Prusak. 1992. Information politics. *Sloan Manage. Rev.* 34: 53-65.

Davis, M.B. 1981. Quaternary history and the stability of forest communities. In *Forest Succession: Concepts and Application*. D.C. West, H.H. Shugart, and D.B. Botkin, eds. Springer-Verlag, New York.

Davis, M.B. 1989. Insights from paleoecology on global change. *Bull. Ecol. Soc. Am.* 70: 222-228.
DeSmedt, W. 1994a. The wolf at the door. *Database Programming & Design* (April): 58-67.
DeSmedt, W. 1994b. CASE and prototyping: mind over model. *Database Programming & Design* (May): 47-49.
Foster, D.R., P.K. Schoonmaker, and S.T.A. Pickett. 1990. Insights from paleoecology to community ecology. *Trends Ecol. Evol.* 5: 119-122.
Freudenberg, W.R. 1992. Nothing recedes like success? Risk analysis and the organizational amplification of risks. *Risk Issues Health Safety* 3: 1-35.
Gaines, S., and J. Roughgarden. 1985. Larval settlement rate. *Proc. Natl. Acad. Sci. USA* 82: 3707-3711.
Geman, D. 1990. Random fields and inverse problems in imaging. *Lecture Notes in Mathematics*. Springer-Verlag, New York.
Geman, D., and B. Gidas. 1991. Image analysis and computer vision. Pp. 9-36 in *Spatial Statistics and Digital Image Analysis*. Panel on Spatial Statistics and Image Processing, National Research Council. National Academy Press, Washington, D.C.
Henderson-Sellers, A. 1990. Predicting generalized ecosystem groups with the NCAR CCM: First steps toward an interactive biosphere. *J. Climate* 3(9): 917-940.
Holling, C.S. 1992. Cross-scale morphology, geometry, and dynamics of ecosystems. *Ecol. Mon.* 62: 447-502.
Jackson, J.B.C. 1994a. Constancy and change of life in the sea. *Philos. Trans. R. Soc. London Ser. B* 344: 55-60.
Jackson, J.B.C. 1994b. Community unity? *Science* 264: 1412-1413.
Kanciruk, P., and M.P. Farrell. 1989. *The Case for Issue-Oriented Information Analysis Centers in Support of the U.S. Global Research Program*. Environmental Sciences Division, Publ. No. 334671189. Oak Ridge National Laboratory, Oak Ridge, Tenn. 30 pp.
Kareiva, P., and M. Anderson. 1988. Spatial aspects of species interactions: the wedding of models and experiments. Pp. 35-50 in *Community Ecology*, A. Hastings, ed. Lecture Notes in Biomathematics 77. Springer-Verlag, Berlin.
Katzenbach, J.R., and D.K. Smith. 1993. *The Wisdom of Teams: Creating the High-Performance Organization*. Harvard Business School Press, Cambridge, Mass.
Kirkpatrick, D. 1993. Making it all worker-friendly. *Fortune* 128(7): 44-53.
Levin, S.A. 1992. The problem of pattern and scale in ecology. *Ecology* 73: 1943-1967.
Lewin, R. 1985. Plant communities resist climate change. *Science* 228: 165-166.
Lincoln, Y.S. (ed.). 1985. *Organization Theory and Inquiry*. Sage, Newbury Park, Calif. 231 pp.
Loehle, C., J. Gladden, and E. Smith. 1990. An assessment methodology for successional systems. I. Null models and the regulatory framework. *Environ. Manage.* 14: 249-258.
Lubchenco, J., A.M. Olson, L.B. Brubaker, S.E. Carpenter, M.M. Holland, S.P. Hubbell, S.A. Levin, J.A. MacMahon, P.A. Matson, J.M. Melillo, H.A. Mooney, C.H. Peterson, H.R. Pulliam, L.A. Real, P.J. Regal, and P.G. Risser. 1991. The Sustainable Biosphere Initiative: An ecological research agenda. *Ecology* 72: 371-442.
May, R.M. 1991. Biodiversity: A fondness for fungi. *Nature* 352: 475-476.
National Research Council (NRC). 1988. *Marine Environmental Monitoring in the Chesapeake Bay*. National Academy Press, Washington, D.C.
National Research Council (NRC). 1991. *Solving the Global Change Puzzle: A U.S. Strategy for Managing Data and Information*. National Academy Press, Washington, D.C.
National Research Council (NRC). 1992. *Combining Information: Statistical Issues and Opportunities for Research*. National Academy Press, Washington, D.C.

National Research Council (NRC). 1995. *Preserving Scientific Data on Our Physical Universe: A New Strategy for Archiving the Nation's Scientific Information Resources.* National Academy Press, Washington, D.C.

Office of Science and Technology Policy (OSTP). 1991. *Policy Statements on Data Management for Global Change Research.* U.S. Global Change Research Program, National Science Foundation, Washington, D.C.

O'Neill, R.V. 1988. Hierarchy theory and global change. Pp. 29-46 in *Scales and Global Change: Spatial and Temporal Variability in Biospheric and Geospheric Processes*, T. Rosswall, R.G. Woodmansee, and P.G. Risser, eds. SCOPE 35. Wiley, New York.

Parsons, T. (ed.). 1947. *Max Weber: The Theory of Social and Economic Organization.* Free Press, New York.

Pennington, W. 1986. Lags in adjustment of vegetation to climate caused by the pace of soil development: Evidence from Britain. *Vegetation* 67: 105-118.

Rasool, S.I., and D.S. Ojima (eds.). 1989. *Pilot Studies for Remote Sensing and Data Management: Report of a Meeting of the IGBP Working Group on Data and Information Systems.* The Royal Swedish Academy of Sciences, Stockholm.

Reason, J. 1990. The contribution of latent human failures to the breakdown of complex systems. *Philos. Trans. R. Soc. London Ser. B* 327: 475-484.

Roughgarden, J., Y. Iwasa, and C. Baxter. 1985. Demographic theory for an open marine population with space-limited recruitment. *Ecology* 66: 54-67.

Roughgarden, J., S.D. Gaines, and S.W. Pacala. 1986. Supply side ecology: The role of physical transport processes. Pp. 491-518 in *Organization of Communities: Past and Present*, J.H.R. Gee and P.S. Giller, eds. Blackwell, Boston, Mass.

Sellers, P.J., F.G. Hall, G. Asrar, D.E. Strebel, and R.E. Murphy. 1992. An overview of the First International Satellite Land Surface Climatology Project (ISLSCP) Field Experiment (FIFE). *J. Geophys. Res.* 97: 18355-18371.

Shugart, H.H., T.M. Smith, and W.M. Post. 1992. The potential for application of individual-based simulation models for assessing the effects of global change. *Ann. Rev. Ecol. Syst.* 23: 15-38.

Steele J.H. 1991. Marine ecosystem dynamics: Competition of scales. *Eco. Res.* 6: 175-183.

Strebel, D.E., J.A. Newcomer, J.P. Ornsby, F.G. Hall, and P.J. Sellers. 1990. The FIFE Information System. *IEEE Transactions on Geoscience and Remote Sensing* 28: 703-710.

Townshend, J.R.G., and S.I. Rasool. 1993. Global change. In *Data for Global Change*, P.S. Glaesar and S. Ruttenberg, eds. CODATA, Paris.

Ustin, S.L., C.A. Wessman, B. Curtiss, E. Kasischke, J. Way, and V.C. Vanderbilt. 1991. Opportunities for using the EOS imaging spectrometers and synthetic aperture radar in ecological models. *Ecology* 72: 1934-1945.

Webb, T., III. 1987. The appearance and disappearance of major vegetational assemblages: Long-term vegetational dynamics in eastern North America. *Vegetation* 69: 177-187.

Weick, K.E. 1976. Educational organizations as loosely coupled systems. *Admin. Sci. Quart.* 21: 1-19.

Weick, K.E. 1982. Management of organizational change among loosely coupled elements. In *Change in Organizations*, P. Goodman, ed. Jossey-Bass, San Francisco, Calif.

Weick, K.E. 1985. Sources of order in underorganized systems: Themes in recent organizational theory. Pp. 106-136 in *Organizational Theory and Inquiry*, Y.S. Lincoln, ed. Sage, Newbury Park, Calif.

Wessman, C.A. 1992. Spatial scales and global change: Bridging the gap from plots to GCM grid cells. *Ann. Rev. Ecol. Syst.* 23: 175-200.

Wiens, J.A. 1989. Spatial scaling in ecology. *Funct. Ecol.* 3: 385-387.

Worster, D. 1977. *Nature's Economy.* Cambridge University Press, New York.

Appendixes

A

Case Study Evaluation Criteria

In order to facilitate fact-finding for the case studies, the committee developed a set of criteria to function as a framework for identifying and analyzing issues involved in interfacing disparate data types. The criteria are intended to assist in learning lessons from past experiences and in developing general principles for future data integration efforts in support of environmental research.

The committee has identified five subject areas to consider in its assessments. These are:

- User needs,
- Study design,
- Data characteristics and quality,
- Data management, and
- Institutional issues.

The specific criteria in each area derive from basic data management and data integration issues. Since these five areas represent a somewhat arbitrary separation of issues, there is some overlap among the specific criteria proposed for each area. In addition, these criteria and their related questions are intended to encompass as wide a range of issues as possible. Therefore, not all of them are equally relevant to every study.

USER NEEDS

Identifying all the users of data and their various needs is vitally important to the successful development and implementation of any data management plan. Given the interdisciplinary nature of much global change research, and the high cost of developing data sets, it is very likely that the user community will include not only existing study participants, but also additional future users. These future users may want to use the data for novel purposes and to interface them with data types beyond those originally envisioned. This requires defining user needs in the broadest sense possible. The term "user needs," as used here, refers to needs to find, evaluate, access, transfer, and/or combine data. It also refers to requirements for manipulating, processing, analyzing, or otherwise working with the data. Finally, it refers to the necessity for users to respond to institutional or cultural constraints, motivations, or pressures. Questions to consider include the following.

Identifying Users

- Was there a clear definition of users and user groups at the inception of the research project?
- Were users at each step of the data path, from initial data collection to final analysis and archiving, clearly defined?

Understanding Users' Requirements

- Were the specific requirements of users at each step of the data path clearly defined?
- Were future potential users' needs predicted and accommodated?
- Were there incompatibilities or conflicts among different user groups?
- Were institutional structures and management mechanisms (committees, working groups) established to identify users' needs and resolve conflicts?
- Did users feel as if their needs were accommodated? If not, why not?

Technical Aspects

- Did the study create specialized algorithms, routines, data management procedures, or database structures to accommodate users' needs? If so, how successful were they?

- Did the study, as originally envisioned, require interfacing disparate databases?
- Were interfacing requirements and issues understood and allowed for?

STUDY DESIGN

There are many design principles related to the conceptual and statistical validity of scientific research. Here the committee considers only those related to interfacing among different data types. Considering potential data interfacing issues in the original study design usually lessens the problems associated with the integration of data. This section considers strictly technical study design issues. Other design issues related to program structure and management are listed below in the section "Institutional Issues." Technical questions to consider include the following.

Conceptual Framework

- Was the study based on an overall conceptual model that described the relationships (both theoretical and functional) among different data types?
- Was the conceptual framework pursued to a level of detail that helped identify data interfacing issues?
- Was the conceptual framework explicitly multidisciplinary and multimedia?

Methodological Issues

- Was the study an interdisciplinary one involving multiple data types?
- Were all relevant disciplines and data types identified at the beginning of the study, or were midcourse adjustments required?
- Were pilot studies performed to assess potential data integration issues and solutions?
- Were data integration issues identified and planned for in the initial phase of the study? If not, at what stage of the study were they considered?
- Were methodological differences among study components that created difficulties in later integration identified at the outset of the study?
- What changes would the participants make in the study design if they had the opportunity to begin over again?

Data Integration

• Did the study design involve using preexisting data? If so, what problems were encountered? Were enough metadata available?

• Were there technical differences among disciplines that created data integration problems, e.g., requirements for different spatial scales or levels of detection?

• What kind of data integration did the study's data analyses require? Were these based on the study's underlying conceptual model and were they allowed for in the study design?

DATA CHARACTERISTICS AND DATA QUALITY

Issues related to data characteristics and quality will arise from a variety of sources. Some studies will combine both new and historical data. Historical data may contain numerous errors, often lack adequate documentation, may be collected or processed inconsistently from place to place or over time, may lack critical quality control information, and may be stored in incompatible formats. New data may represent a wide variety of data types, as well as spatial and temporal scales. Data volumes are typically large for climate change studies. Quality control is of paramount importance, since errors can occur not only at the time of collection and initial processing, but also at any time the data are accessed and used. Questions to consider include the following.

Data Characteristics

• Were data characteristics sufficiently documented in the metadata? If not, how difficult was it to find needed information about the data?

Quality Control

• If historical data were used, what quality control problems were encountered and how were they resolved?

• Were potentially problematic data characteristics known beforehand or discovered in the data integration process?

• How were differences in data quality among data sets handled?

• Were data quality procedures considered an integral part of data integration?

• How were data verified and validated?

Data Integration

- What specific data characteristics created data interfacing problems?
- What was the source(s) of these problems?
- Were there data formatting or quality standards that proved useful?
- What lessons were learned that would be applicable to other studies?

DATA MANAGEMENT

Data management refers to the provisions for handling the data at each step of the data path, from initial study design, through data collection, accessing, and analysis, to final reporting and archiving. It refers not only to specific technical procedures, but also to the overall plan for ensuring the original quality of the data and preventing their degradation over time. Data management plans should include organizational plans that specify data management functions and who has responsibility for data quality at each step of the data path. Relevant questions include the following.

Up-Front Planning

- Was there an overall data management plan that supported the data integration process?
- What provisions were made for data access, retrieval, and manipulation?
- Were data management procedures designed to relate directly to technical issues involved in data integration?
- Were quality control issues considered in all data management procedures?
- Were archival needs considered at the beginning of the study?

Data Management Procedures

- Were specific database tools developed to aid the database interfacing process?
- Are there readily identifiable authorized versions of the different data sets? If so, how are these maintained?
- What provisions were made to make metadata available to users?
- Did data management requirements related to database interfacing add to project overhead?

- Did data integration directly benefit project participants?
- How accessible were the data?
- Were there any restrictions on use of the data? If so, what was the source of these restrictions?

Planning for the Future

- Did data management procedures and systems explicitly consider future potential needs?
- What arrangements were made for archiving the data for future uses?
- Where are the data now and are they easily accessible? Are metadata readily available for future users?
- How easy would it be to transfer existing data to different database systems?
- What changes would the participants make in the data management plan if they had the opportunity to begin again?

INSTITUTIONAL ISSUES

Institutional issues often have an overriding influence on the success of data integration efforts, yet they can be difficult to identify and resolve. These issues arise, for example, from differences in agency missions and mandates, from funding restrictions, from differences in time horizons and constituencies, and from differences in organizational cultures. Relevant questions include the following.

Participants

- Who were the key participants and what were their roles, responsibilities, and authority?
- What was the nature of the key participants, e.g., private, governmental?
- Were key players or data sources missing from the study?
- Did any participants place special conditions on their participation and/or on access to data, e.g., proprietary data?

Organization and Management

- What was the project's management structure, especially with regard to database interfacing? Was there a lead entity?
- Did the study's organizational structure support or impede database interfacing?
- What arrangements were made among the participants with regard to database interfacing? Were these formal or informal?
- What was the decision-making process, again especially with regard to database interfacing?
- What kinds of arrangements were made for acquiring data from other organizations?
- Was adequate funding available and committed for the duration of the study?
- Was there a long-term commitment to database updating and other maintenance?
- Who can access the data now and are there any restrictions on this?
- What agency, if any, was given responsibility for long-term management and maintenance of the data?

Data Integration

- Did all participants agree with the need for data integration?
- What mechanisms were established for cooperation and data integration? Were any of these novel?
- Were potential conflicts and disagreements clearly identified and negotiated at the beginning of the study?
- Did agency missions, mandates, and policies restrict participation or otherwise impede database interfacing?
- Did existing data management practices impede data integration?

B

List of Abbreviations and Acronyms

AISC	Assessment and Information Services Center
ANC	Acid neutralizing capacity
ARM	Atmospheric Radiation Measurement (Program)
AVHRR	Advanced Very High Resolution Radiometer
BOREAS	Boreal Ecosystem-Atmosphere Study
CalCOFI	California Cooperative Oceanic Fisheries Investigation
CDIAC	Carbon Dioxide Information Analysis Center
CD-ROM	Compact Disk-Read Only Memory
CEES	Committee on Earth and Environmental Sciences
CIAD	Climate Impact Assessment Division of NOAA's Assessment and Information Services Center
CODATA	Committee on Data for Science and Technology
DAAC	Distributed Active Archive Center
DBMS	Data Base Management System
DDRP	Direct/Delayed Response Project
DOE	Department of Energy
EPA	Environmental Protection Agency
EROS	Earth Resources Observing System
ERP	Episodic Response Project
ESD	Environmental Sciences Division
FCMA	Fisheries Conservation and Management Act
FIFE	First ISLSCP Field Experiment
FSDB	Forest Science Data Bank
GIS	Geographic Information System

APPENDIX B

GOES	Geostationary Operational Environmental Satellite
ICSU	International Council of Scientific Unions
IGBP	International Geosphere-Biosphere Program
ILWAS	Integrated Lake-Watershed Acidification Study
ISLSCP	International Satellite Land Surface Climatology Project
LTER	Long-Term Ecological Research (Program)
MAGIC	Model of Acidification of Groundwater in Catchments
NADP	Nicotinamide adenine dinucleotide phosphate
NAPAP	National Acid Precipitation Assessment Program
NASA	National Aeronautics and Space Administration
NDVI	Normalized Data Vegetation Index
NMFS	National Marine Fisheries Service
NOAA	National Oceanic and Atmospheric Administration
NRC	National Research Council
NSF	National Science Foundation
NSWS	National Surface Water Survey
ORNL	Oak Ridge National Laboratory
OSTP	Office of Science and Technology Policy
PC	Personal computer
QA	Quality assurance
QC	Quality control
USAID	U.S. Agency for International Development
USNC	U.S. National Committee
WAIS	Wide Area Information Servers
WMO	World Meteorological Organization